Welton da Cruz Martins

Principles of elasticity and variable constant theory

3rd edition

São Luís - MA
2022

Author
Welton da Cruz Martins
Graduated in Chemistry, Specialist in Higher Education Teaching, Professor, Researcher, Inventor and Writer.

Guidance and review:
Davina Camelo Chaves
PhD in Chemistry from the Federal University of Ceará. Professor of the EBTT of IFMA, researcher, writer and general director of the Zé Doca Campus.

Willinilson Wild dos Santos Lago
Degree in Physics from the Federal University of Maranhão. Professor at IFMA – Campus Zé Doca IFMA.

Mathematical review and collaboration:
Rruydeiglan de Sousa Lima
Graduated in Physics, Master in Physics, Prof. EBTT of IFMA – Campus Zé Doca, Researcher and also collaborator of this work.

Rodson Regi de Sousa Correia
Mathematician and Professor Esp. of IFMA – Campus Zé Doca

Spelling review:
Raimundo Nonato Marques do Nascimento
Philosopher and master in theology, Prof. EBTT of IFMA – Campus Zé Doca, writer and researcher.

Cover and graphics:
Dhemeson Silva
Graduated in Chemistry, researcher, inventor and writer.

Collaborating Researchers:
Joaldo da Silva Lopes
Industrial Chemist and Professor at IFMA – Caxias Campus
Marcos Fernandes Silva
Graduating in Chemistry at IFMA – Campus Zé Doca
Joseph of Arimatéia Cunha
Mechanical Engineer and Prof. Esp. at IFMA – Campus Zé Doca

São Luís - MA
2022

Catalog Data

Martins, W. C.; 1992;

Title: Principles of elasticity and variable state theory/ Welton da Cruz Martins. – Zé Doca, MA. 3rd edition. 2022.

143 p.; 21 cm.

ISBN: 978-85-921379-0-8

Elasticity. 2. Elastic and dialectical-thriving systems. I. Title

CDU: 620.172.22

Dedication

To God,

To my parents, Gilberto and Maria de Fátima Martins, because without them I don't see how this would be possible,

To my partner, Flavia Matos, who supported and helped me a lot,

To my Sister, for her good words,

To Professor Davina Chaves, for believing in me, encouraging and saying move on, when my faith in work weakened.

To Professor Wild Lago, for listening to me and discussing from the beginning,

To Professor Ruydeiglan Lima, for his great help and contribution.

To IFMA – Campus Zé Doca, for contributing to this work,

To all my teachers, especially teacher Lene and Lidiana

To all friends,

To you who purchased this work.

Summary

Preamble

Elasticity is not just a mathematical justification, it is an entity, acting at different levels and systems, responsible for everything. The elastic constants must be measured considering not only the body but also the temperature it is in, since there are variations in the speed and strength of the response of the materials as a function of temperature. Just as a liter of water is not the same volume at different temperatures and does not have the same physical characteristics, other materials, even if solid, also do so, but each material in its uniqueness.

Chemical reactions may not be the same, have the same products, concentrations and numerical variables, at different points in space-time. Elastic behavior can be different under different conditions of gravity and space-time. The entire structure of matter and response to demands is dependent on its energetic and bodily conditions.

Note that these relationships are also manifested through waves and that these waves (caused by vibrations generated by disturbances) can destroy solid bodies by requiring a demand for intensity at a given limit frequency capable of causing them to suffer fractures, which probably may have happened in the famous story of the book of Exodus narrated in the Bible in which the walls of the city of Jericho fell, by having a large army marching and making sounds; The phenomenon known today as resonance may have been responsible for the fall of the walls.

In this book, a new model of elastic body is proposed, which has as its main characteristic the ability to have its elastic constant modified to meet the demands of related loads. Whose commercialization is protected by patent with the INPI, however, is free and the research of others on the subject is encouraged.

The creation of these new springs makes Robert Hooke's principle of invariability of the elastic constant at the same temperature relative. A perspective capable not only of introducing a new branch of research to the sciences, but of satisfying a range of human needs in what tends to elastic demands.

If this spring is used in vehicles, it can be loaded with full load and still remain with the same damping capacity and elasticity. Because it can self-adjust the load requests, enabling the best elastic or dissipative response possible.

It is in the author's interest to carry out experiments that confront his data, analyze his perspectives and propose improvements. For only in this way, with impersonal, critical, logical, sincere and exhaustive investigation, can a science of value be built.

This book is an essay, which originated a new product, at the beginning of my academic life. I kindly ask the reader, if possible, for your opinion in the author's e-mail.

What is elasticity

Without elasticity there would be no life, no movement or chemical bonds. The universe would be impenetrable or extremely unstable, there would be no time and matter, and the existence of anything along the lines of actual matter would be impractical.

The term elasticity refers to the characteristic of bodies restoring their original shape after the removal of a charge demand, that is, the ability of a being to stretch itself before a force and restore the original form after the action of this force. We use the term "being" here because, as we will see, all things admit an elastic behavior that only exists from physical interactions.

One of the first scientists to research elastic bodies and propose an explanation was the Englishman Robert Hooke in approximately 1662, from whom all other elastic theories originated. His postulate can be summarized as follows:

"An elastic body is any body that has the ability to restore its shape after the discharge of a force."

This behavior can be described simply by , where F is the applied force, k is the elastic constant, and x is the deformation suffered. However, it is not only solid materials that have elasticity associated with their kinetic and mechanical behavior described by this equation or relationships with it, but gases, liquids and the various conformational forms of energy, matter and space-time. $F = k.x$

Hooke described a simple and objective way to understand why some bodies, especially those with a high elastic constant (k), are more rigid and less elastic, while bodies with a lower elastic constant are more flexible. It was the beginning of the understanding of a totally new area of mechanics, which over time would extend to the quantum equations of the behavior of electrons, chemical species, gases, liquids, solids, among others. Propagated and improved by several scientists, such as Young (elasticity of solids), Boltzmann (behavior of gases), Einstein (Browlian motion and deformation of space-time) and many others.

The elastic behavior of bodies from other states of matter, such as plasma, fermionic, and others, is less well known, although it is known to exist in different conformations and quantities. Studies with elasticity at the nanotechnological level and organometallic alloys have shown promise and are widely studied at a closed level, with several elastic entities used, not just materials.

In this perspective, entities are classified as corporeal, those formed by matter at the macroscopic level or similar, quantum entities, formed at the quantum level or with a purely energetic and interactional character, abstract entities, at the level of the field of ideas, and anomalous entities, which do not fit into any other classification.

In this way, it is possible to find interstices between classes of elastic entities such as quantum/corporeal with ambiguous, double or triple behavior, as is the case of some electrically charged nanoparticles immersed in fluids that admit both quantum and corporeal behavior, similar to Browlian motion.

The union or interaction of more than one elastic entity is called an elastic system, being of two types and two classes, respectively, harmonic or disharmonious for types whose interaction favors or disfavors elastic behavior. As for classes, we call Hooke's system or simple, when there are regular interactions as predicted by Robert Hooke and dialectical-powerful systems, with interactions of the type described in chapter 5 of this book, whose degree of elasticity depends on intrinsic characteristics of those involved and the strength of the contact that we will see in chapter 2.

Therefore, we first evaluate some classes of elastic entities. It is important at this stage that you seek to understand the differences and similarities in the reading, it is worth remembering that most contributions to physics try to unify concepts after several theories and experiments, but for elasticity there was first a unifying point (Robert Hooke), from where all the other laws came, so be careful, because some statements that we will make implied new concepts and positions, little detailed in the literature.

ELASTICITY OF MATERIALS

With regard to corporeal elastic entities, that is, materials, we can analyze their behavior from at least two perspectives, first, macro-mechanistic from the rheological point of view, then, micro-mechanistic, referring to the microstructures and/or molecules responsible for the characteristics presented in the macro-mechanistic analysis. Both are equally essential for the completeness of the understanding of elasticity.

This is because intermolecular bonds act together with a body in order to give it elastic properties. In fact, these are responsible for much of the elastic behaviors we observe in the macroscopic world.

Although, chemical bonds (covalent, metallic and ionic) and the structures of unit cells/molecules have the greatest contribution, since they are the ones, in most cases, that will allow or not the work to be carried out on the body, which is possible through the intermolecular forces that keep them cohesive and firm.

Even though there are other kinds of elastic behaviors that we must consider, which may depend entirely on intermolecular forces, as in certain non-Newtonian fluids, we consider a priori simple and usual, solid materials.

When it comes to elasticity of materials, it is evident, among the various existing classes, some of which are described in chapter three. In general, they present behaviors whose elastic constant for a given class varies at regular intervals, as shown in Table 1. This occurs due to the arrangements of the molecular structures or unit cells of the same and the similarity of classical — a set of characteristics that conditions groups of materials in a given class —, due to this first characteristic, some materials may have two moduli of elasticity or more, one for each position of the unit cell or molecular arrangement that is requested.

TABLE 1. Modulus of elasticity of some materials.

Material	Elastic Modulus (GPA)	Unit cell
Rubber	<<1	Absent
Aluminium	69	CFC
Titanium	107	H
Tungsten	407	CCC
Glass	48 - 83	Amorphous
Diamond	1000	CFC

Source: adapted by the author himself

When the same body has two elastic moduli or more, these are verified by tests in different positions of the unit cells of the same and the body is called anisotropic, see that this placement generally used to describe geometrically non-uniform things is used here to express a lack of structural uniformity from the molecular point of view and physical from the point of view of the distribution of energy by a demand applied to the body.

Materials whose modulus of elasticity can be considered the same, regardless of the position of the body during the test (test), are called isotropic. Figure 1 shows the cells and systems of Bravais, where it can be seen that there is a possibility of anisotropy for bodies with certain constitutional systems and cells.

FIGURE 1. Bravis cells and systems according to crystallinity theory.

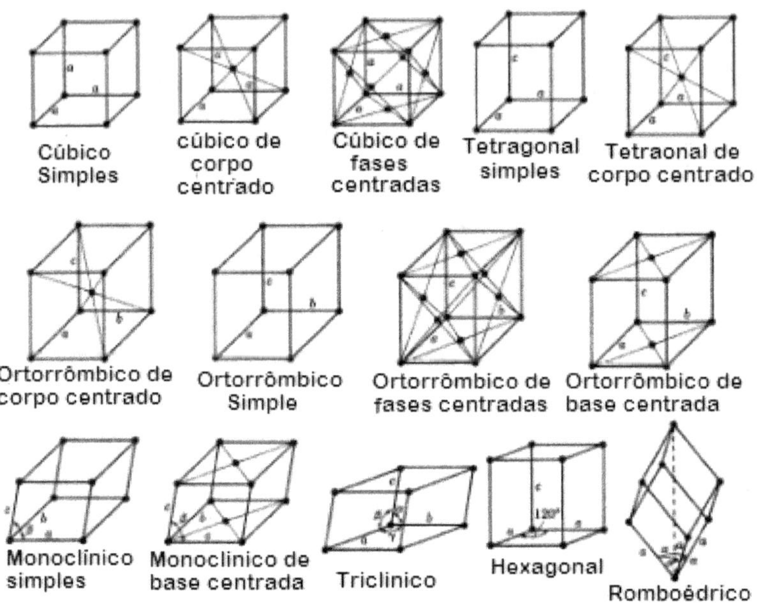

Source: adapted by the author himself

These cells or reticula are generally described by three cell factors and three angles, such as the crystalline systems shown in table 2, where it can be seen that for some systems anisotropy is an existential characteristic, imposed by the angles and factors of the cell, at its core or when in a system, which depend on chemical and physical characteristics of the constitutional species.

When a body is elastic, it means that its atoms can move in degrees of freedom without breaking chemical bonds.

Table 2. Parameters and angles of crystalline systems

System	Parameters	Angles	Cells
Cubic	a =b=c	a = b = y = 90°	CS, CCC e CFC
Tetragonal	a=b≠c	a = b = y = 90°	TS and TCC
Orthorhombic	A≠B≠C	a = b = y = 90°	OCC, OS, OFC and OBC
Rhombohedral	a=b≠c	a ≠ b ≠ y ≠ 90°	R
Hexagonal	a=b=c	a = b =90°, y = 120°	H
Monoclinic	A≠B≠C	a = y =90°, b < 90°	MS and MBC
Triclinic	A≠B≠C	a ≠ b ≠ y ≠ 90°	T

Source: adapted by the author himself

Molecular arrangements, in the case of non-crystalline compounds, are responsible for classifying them into bands – in most cases – since the organization of molecules, as in the case of polymers, can be random and, therefore, each class of material will have a minimum/maximum limit of nonlinear elastic modulus. This is what happens in most elastomers, plastics, and even some alloys.

Elastomers are polymers that have a high elastic capacity, they are usually called rubbers and elastics. These polymers have long, coiled chains with remarkable intermolecular bonds. When elastomers undergo a pull, the molecular chains unravel and the intermolecular bonds are momentarily overcome by the pulling force, which are restored when the charge is removed, returning almost completely to the previous position.

In general, the properties of materials depend on the energy of the chemical bonds, the geometric arrangements of the species that constitute them, and the energy in transit. For this reason, temperature influences the strength and elasticity of materials.

When an elastic material is subjected to a load demand capable of stretching it, the total external energy supplied to the body is transformed into at least three distinct forms, where a small fraction is lost to friction – usually in the form of heat – another is transformed into kinetic energy and the third part into elastic potential energy. The latter will be responsible for restoring the original shape of the body, transformed into kinetic energy when the traction force is removed, which does not always happen in a precise or integral way, since the perfectly elastic behavior is until then, only theoretical.

If we imagine a perfectly elastic system, for example, gases at a given temperature in an isolated system or even a shuttlecock that bounces eternally on the ground, because it does not lose energy we would have to consider it a perfect system. However, understanding the elastic system as a thermodynamic process, the Carnot cycle and the second law of thermodynamics described by Boltzmann make the reality of this type of process unfeasible. Today it is admissible, only on a few occasions at the quantum level, where even Lavoisier's principle of conservation of energy does not seem to apply. Thus, it is obliged to admit, according to Carnot's understanding, that:

"No interaction can be perfectly elastic at the macroscopic level."

The behavior described above is typical of any elastic body, whether it is a piece of glass, an elastomer or a strand of hair, in fact what changes from one body to another, in terms of Hooke's law is the elastic constant, but all work consists of the admission of losing energy beyond the actual expenditure of its expenditure.

Thus, glass is less elastic than rubber because it admits the performance of little elastic work, becoming extremely brittle, having a high elastic constant and in the case of elastomer, with a small value of the constant it becomes very elastic in relation to glass, although there are elastomers with considerable hardness and toughness. Remember that hardness is the ability to take pieces by scratches and not to resist impacts, which is toughness.

The strength of materials depends, among other things, on their ductility, that is, the ability of atoms to slide over each other. This capacity requires elasticity as well as moldability. However, not every ductile body will be essentially very elastic, because although it is necessary for some materials such as elastomers to be elastic, many elastic processes do not need this characteristic as much as concrete and glass, although it is notable up to a certain limit of ductility, that the more ductile the more elastic the materials tend to be, when intermolecular forces act together.

Much of the elasticity depends on the elastic limit of the materials. This limit is the point at which the stress or tensile induces permanent deformation, which when it occurs, is indicative that the ductility is greater than the strength of the interatomic or intermolecular interactions in the material or even, but commonly, that bonds have been broken. Exceeding this point causes the material to lose much of its physical properties, usually becoming weaker and more brittle.

GRAPH 1. Analysis of the behavior of materials.

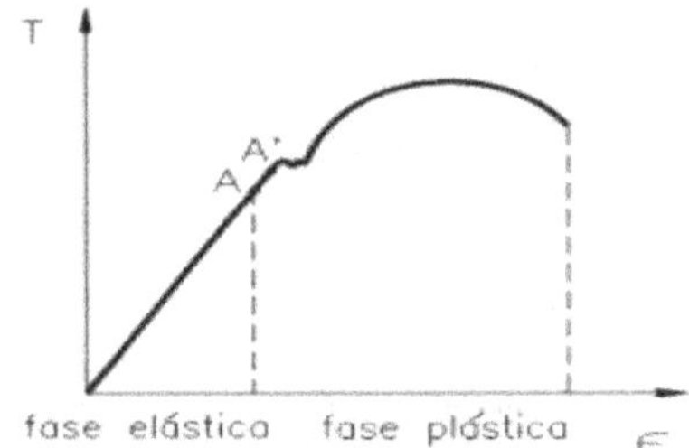

Source: Mascia 2006.

In steels and most materials, the elastic limit corresponds to the proportionality limit. That is, to the extent that the force continues to be proportional to the deformation, in elastomers the limit of proportionality continues even after the limit of proportionality. Hooke's law "only holds" up to the limit of proportionality of materials, after which calculations such as Cauchy and others that we will see in chapter 2 are used for certain materials, however, after passing the elastic limit the body is fatigued.

Graph 1 shows a linearly elastic material, undergoing plasticity process, however, there are non-linearly elastic materials, where A corresponds to the elastic safety limit and A' is the starting point of plastic deformation, from which the body can be considered fatigued, losing its initial properties and being conducive to continued cutting the effort.

GRAPH 2. Behavior of elastic solids in terms of linearity.

Source: Dalcin 2007

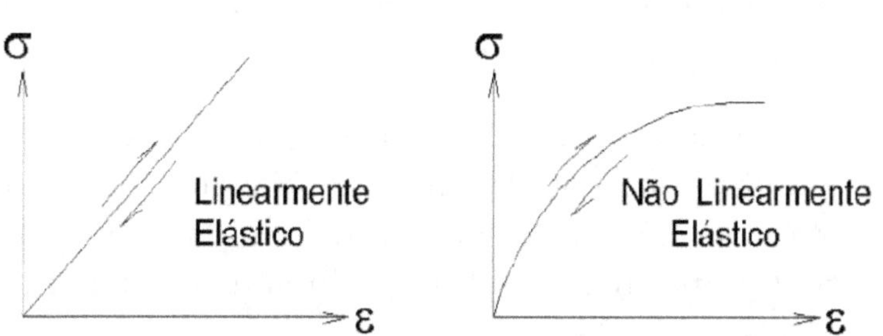

Fatigue and subsequent disruption occur due to the breaking of the chemical bonds between the species at the points where the greatest demands are concentrated, it is this point that actually "appears" in the plastic zone, the cell is destroyed and the system undone, the same occurs with the molecular arrangements for non-crystalline materials.

In graph 2, linearly elastic and non-linearly elastic bodies are observed, which do not change their elastic constant before the plastic phase, observed in graph 1. However, the graph in the form of a parabola, in order to have its constant calculated, sometimes uses a mathematical artifice, which we will see in detail in chapter 2 along with the equations necessary for mechanical tests. An example of a non-linearly elastic material is concrete and latex sex condoms.

In this sense, all bodies of matter are elastic, the very capacity to vibrate can already be considered a manifestation of this character.

The tolerance of materials to a load demand by means of elasticity is such a common means of energy transport, that sometimes we do not perceive it, however, without which nature and the synthetic, whether or not it is linearly elastic, with small or large Young's modulus, would simply be impractical.

From a general view it can be said that the entire planet vibrates, and is therefore elastic, from the response of the ground to a drop of water that falls on it, to a wave that propagates in a fluid medium to a rock to the tremor of a tectonic plate.

Without elasticity, the contact between two bodies – shock – would be essentially destructive. Physical restitution would not be possible, after a demand for force that resulted in deformation, it would tend to be permanent in a certain way if the ductility did not allow breakage.

BONDING ELASTIC QUANTS

Today we know that the molecules that make up everything around us, including ourselves, are in a vibrational state of constant motion, where the angle, length, and even the positions of chemical bonds are constantly changing. These movements are unique to each molecule and can be observed mainly in infrared spectra, such as the one presented in spectrum 1 of a species developed at IFMA – Campus Zé Doca, with the formula $C_nH_yO_z$.

The presence of elasticity in chemical bonds is evidenced in the natural frequency of bond vibrations, given by (eq.1), where the elastic constant (K) is present and varies from one bond to another. The μ is the reduced mass, which corresponds to the ratio between the sum

20

and production of the mass of all atoms in grams, and the wavenumber ($\frac{1}{\lambda}$cm⁻¹), which is the inverse of wavelength λ.

$$\frac{1}{\lambda} = \frac{1}{2\pi c} \cdot \sqrt{\frac{K}{\mu}} \qquad \text{(Eq.1)}$$

When a molecule emits or absorbs a given frequency of electromagnetic wave in the infrared region, it means that it is possible to construct infrared spectra that characterize the bond length between the species that constitute it and, as this length is unique for each bond, it is therefore possible to identify the species that are connected. The main elastic behaviors of the bonds of the molecule under analysis are exemplified after the spectrum, in which the vibrations of the bonds of the compound under analysis are seen.

SPECTRUM 1. Infrared spectrum.

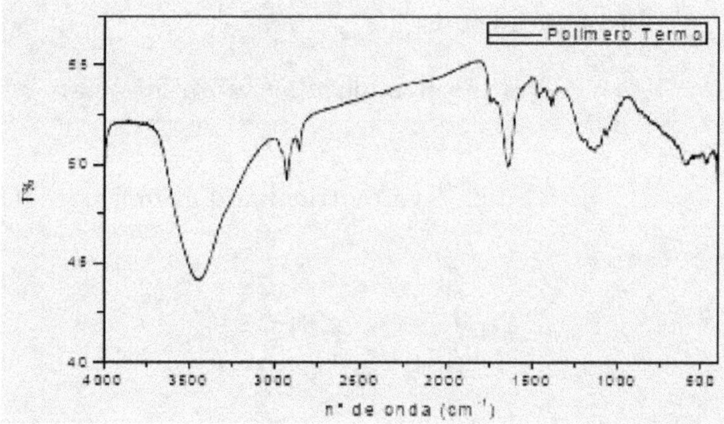

Source: author himself

The intensification of vibration, generated by heat, for example, makes the bonds weaker and weaker, the greater the energy in transit, so that the elastic constant of the system as a whole mutes with temperate intervals.

21

However, the presence of other more electropositive or electronegative atoms and even the solvent can displace the peaks of spectrum 1 slightly, due to a small change in the elastic constant, although very similar to the standard average.

Even if absolute zero is reached, it is believed that the interatomic distances — that is, the bonds — would continue to vary even modestly. This inconstancy of length is the reason why the average length of a bond is usually calculated in X-ray measurements, for example. Another important characteristic is that the elasticity constant undergoes slight changes as a function of partial charges in different molecules, which produces singularities in the spectra and/or change in the bond length, so it is customary to work with wavelength ranges for certain bonds.

The main molecular vibrations can be observed below:

Stretch: symmetrical and asymmetrical

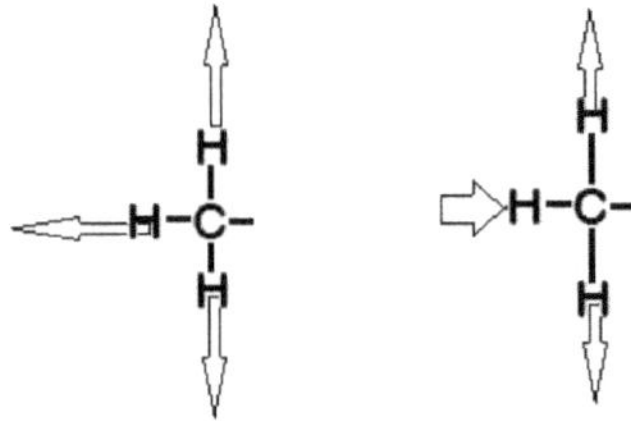

Folding: Off-plane and on-plane

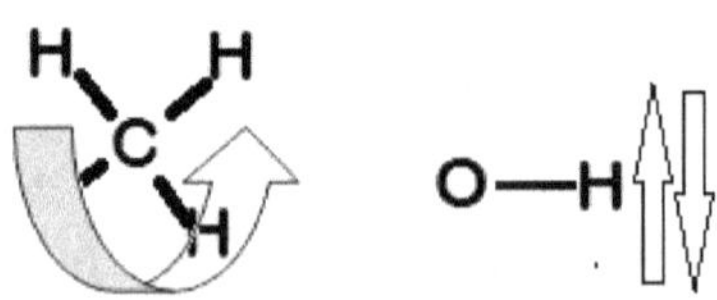

There are several types of vibration, of which only four were mentioned, however, for each type there are many possible frequencies, which is the identity for each molecule or chemical species.

Table 3 shows some of the bond lengths that can be obtained from spectrum 1 and the species that form them. These bond lengths are explained by the wavelength equation (in wavenumber $1/\lambda$) adapted with the Hooke equation.

TABLE 3. Bond lengths and shape of absorption bands.

Connection type	Wavenumber cm-1	Band form
C-OH	~3500	Strong and wide
C-H	2800 to 2900	Weak duo
C-C	900 to 1300	Narrow
O-H	3.400 - 3.200	Wide and strong

Source: author himself

However, in the macroscopic world, the observation of the molecular vibrations of a body can be evidenced through temperature, so when a solid or liquid body is heated the tendency is to increase the interatomic distance and consequently, increase the size of the body (there are exceptions), so it is not incorrect to say that there is an elastic response of the materials to temperature variation.

This temperature variation causes different responses according to the nature of the body and type of material, so a given material can become more elastic when heated, assuming new elastic modulus, or even become more fragile and not very elastic, with large elastic modulus.

The phenomenon of elasticity at the quantum level can be applied to various situations. What is worth considering from the interaction between nuclear hadrons and quarks, as well as the collision between photos and electros of the photoelectric effect, since for some situations the energy is completely conserved and in others it disrespects Lavoisier, where the energy simply disappears.

ELASTICITY IN THERMODYNAMICS

When elasticity is extended to liquids and gases, this behavior is often called viscoelastic. The main difference is that the molecules of both can permanently change places in space and are in constant "flight" motion, which makes their elastic behavior quite unique.

In general, it is usually said that an expanding gas, which returns to its initial volume after the removal of the energy that caused its expansion, has viscoelastic behavior, since it admits both viscous and elastic behavior simultaneously.

FIGURE 2. The behavior of a gas under the action of a force.

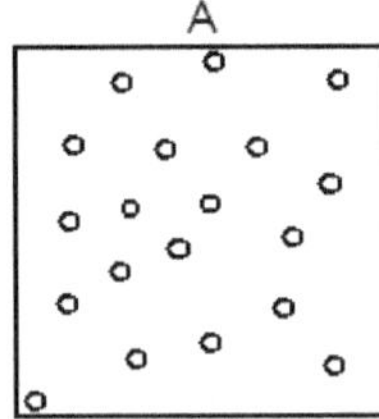
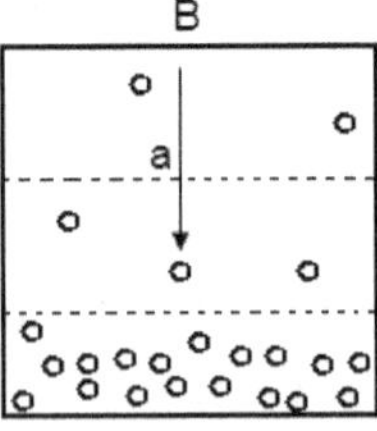
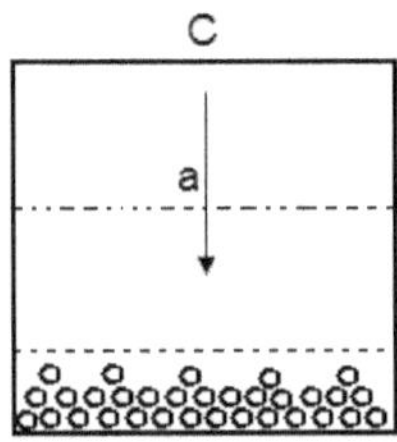

Source: author himself

Thus, when we observe a thermodynamic system from a micromolecular view as observed (figure 2), it can be said that a system that changes the distribution of matter as a function of gravity, in A there is a system free of gravitational action and the shock between the particles is solely a function of temperature, its permanence or not together, depends on the loss or gain of energy associated with this shock and the repulsive attractive forces of the bond if effected.

In B, on the other hand, there would be a system in medium gravity, but the elasticity of the shocks (they are supposed to be perfectly elastic) prevails as a function not only of the temperature, but also of the weight of the atoms. In C, in a hypergravity situation, an open system loses heat and the shock between the molecules can become weak enough for it to change phase. Where is the elasticity?

If atoms did not maintain their elastic bodily structure, at low temperatures atoms would break apart when they were compressed against each other. The question is, how much elasticity must an atom have in order not to break and under what conditions can this happen? The answer lies in nuclear fission, which we will not contemplate here.

It can be said that most things in the universe have elastic behavior. For example, when atoms collide with the clash between species it can be considered (completely) elastic, since they restore their original shape and the shock does not produce significant energy loss when two gas molecules collide (without generating a third). This can be observed in the calculation of the kinetic energy of the molecules (eq.2)

$$E_c = \frac{m.\left(\frac{3RT}{M}\right)^{1/2}}{2} \qquad (Eq.2)$$

Where m is the mass of the molecule, M is the number of moles, R is the constant of the gases, and T is the temperature. Which considers the same mass spheres by the law of ideal gases (Eq.3) and real gases, given by the Van der Waals equation (Eq. 4).

$$PV - NRT \qquad (Eq.3)$$

$$P = \frac{nRT}{V-nb} - \frac{n^2a}{V^2} \qquad (Eq.4)$$

The latter considers the intermolecular forces between two particles to verify whether they will be together or not after the shock, so that the elasticity of the shock depends on the forces of attraction (first side of the equation) and repulsion (second side of the equation). Thus, the dependence on the occurrence of a chemical bond also depends on the elastic character of the entities. The same principle works for calculating the frequency of molecular collisions Eq. 5.

$$ZAB \ (Eq.5) = N_A^2 . \pi (r_A + r_B)^2 \left[\frac{8RT}{\pi}\left(\frac{1}{Ma} + \frac{1}{Mb}\right)\right]^{1/2} [A][B]$$

Where NA is Avogadro's number, r is the radii of the species, M is the moles, A and B are the mass of the colliding molecules. Although there are corrections to the experimental and theoretical model, since not every shock is perfectly elastic, there are other forces of interference in this behavior, as well as limitations of the laws and

26

descriptive equations. The model shows how the energy of the elastic shock is satisfactorily maintained within the system.

When the repulsive forces are stronger than the attractive ones and even in the opposite situation, i.e., when the shock results in a chemical bond – and thus the frequency of collisions decreases – the elastic behavior is maintained although the shock can no longer be considered fully elastic. See figure 4. However, in most cases when the attractive forces, according to the Van der Waals equation, are greater than the repulsive ones, there is only a phase change, because the shock does not have the activation energy necessary to form chemical bonds, only intermolecular bonds.

FIGURE 3. Chemical reaction occurring due to attractive

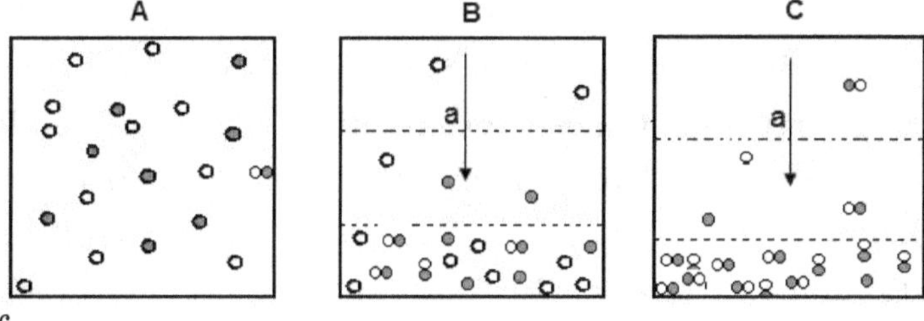

forces.

Source: author himself

The Van der Waals equation acts in the field of Van der Waals attractive forces, dipole-dipole hydrogen bonds, etc. when the shock results in bonding, another elastic behavior acts from there, pertinent to the bond itself, as seen in the previous topic.

Under typical conditions of thermodynamics, the heat itself, responsible for most molecular shocks, and the very

27

order of the system, said entropy, depend on the incredible elastic property of the molecules. Note that the amount of heat in a body depends on the degree of agitation of the particles that constitute it, and this agitation is both intermolecular and extramolecular, when the molecules themselves move and collide with each other.

Even when two solid bodies collide, both respond with an elastic demand. Note that if their elastic potential energy were zero, the shock would immediately result in a permanent deformation capable of decharacterizing both or their complete crumbling, which depends on their ductility. However, the energy dissipates in the connections and structures of the body as a wave, causing them to vibrate, before breaking when the shock exceeds the resistance of the material.

Without shock, elastic bands and molecules would suffer severe damage, not one atom of the same element would be the same as another and matter would be doomed to destruction and reconstitution continuously and on the other hand it would not even be possible, since it depends on the dynamic balance of opposing forces and/or with different signs for its structure to be conceived, which in itself already corresponds to an elastic demand.

THE ELASTIC BEHAVIOR OF RESONANCE

When a body vibrates as a function of a wave — sound or electromagnetic — produced by another, we call it resonance and it is a classic example of the interaction of the shock of the atoms of a fluid with a solid body, making them vibrate at the same frequency, amplifying the wave back to the direction it came.

It is for this reason that instruments such as Clarinet and Saxophone have a wooden reed in the mouthpiece, which vibrates with the air current blown by the musician, producing a wave that resonates and is amplified as it travels through the body of the instrument. The same happens with the guitar, whose string movement emits waves that are resonated inside the instrument's case, amplifying itself.

In the so-called nuclear magnetic resonance machines, the amplification is of an electromagnetic wave, and depends on the ability of the electrons to move from one sublevel to another, absorbing photons, so that the change in field is modified and then the atom is identified, however, a priori, the resonance occurs in the nucleus and has repercussions in the electrosphere.

Equation 6 describes the resonant behavior of a body wave, and resonance only occurs when both are integral functions of a periodic interval. Where T is the frequency and P is the period. a_n ou $b_n \neq 0$

$$F(u) = \frac{a_0}{2} + \sum_{n=1}^{\infty} \frac{2n\pi T}{P} + b_n.\, sen\, \frac{2n\pi T}{P} \qquad (Eq.6)$$

The elasticity of materials associated with a resonance is of great importance in the multiple applications that exist. Ranging from the breaking of a crystal cup to huge concrete bridges by simple, seemingly harmless frequencies of sound waves. To get an idea, the simple march of soldiers can emit waves at a frequency sufficient to collapse walls and bridges of stone and

29

concrete, which occurs when the elastic demand required by the wavelength and intensity is greater than the capacity of the chemical bonds of the material to resist the effort.

ELASTICITY OF SPACE-TIME

Space was considered by many to be nothing, but when Albert Einstein published the theory of general relativity, it can be seen that we are far from truly registering nothing. Space is a kind of "elastic fluid" and for Einstein inseparable from time, so what you see is a space-time mesh.

Einstein proved that space-time can be curved when a sufficiently large force is discharged onto it, such as the mass of a star. This was proven, among other ways, when a star was observed from Ceará – BR and other parts of the world that could not be seen during the eclipse if its light was not bent by the enormous deformation that the sun causes in space-time.

Attention is drawn at this point to the following fact, one does not see many deformations in space without there being a body of matter that produces it, in other words for there to be a gravitational field there must be a body of matter, – on some occasions it is possible that space-time curves in the absence of matter – however, there is still no admissible observational evidence beyond what is called dark matter and Einstein's admitted cosmological constant.

The fact is that at the very least it must be admitted that space recomposes itself in a viscoelastic way, otherwise the orbits of the planets would be true craters where a doughnut-shaped gravitational field line would attract bodies to this orbit, which when moving would cause grooves in space-time, which would be evident by the increase in the density of bodies in these orbital regions. Gravitational waves are impossible, different from what is observed today.

One of Einstein's best-known equations that shows with simplicity the contraction of space (Eq. 7) that evidences this ability by showing that space is distorted when a body approaches the speed of light, but returns to normal conditions when the body ceases the observed motion. It is still possible to see the same elastic phenomenon when observing what happens to time at high speeds, in the equation of time contraction (Eq.8) in which time can stretch according to the speed of the body that performs a displacement.

$$L = \sqrt{1 - \frac{v^2}{c^2}} . L_p \ \text{Eq.}\,7$$

$$\Delta t_f = \frac{\Delta t_i}{\sqrt{\left(1 - \frac{v^2}{c^2}\right)}} \text{Eq.}\,8$$

These equations show that even what was thought to be nothing depends on an elastic equilibrium, to get an idea of the importance of space-time having elastic behavior, Einstein himself postulated that there is nothing that differentiates an accelerated body from a body under the action of gravity for an observer inside the system, in fact because both cause a certain distortion in the homogeneity of space-time. Now see that if space were not elastic, there would be cracks or ditches in it, coming from accelerated bodies or their course.

As time is "inseparable" from space, it can also be stretched or contracted, an analysis of this malleability of space-time is observed in the paradigm of twins, in which one is subjected to an interplanetary travel at the speed of light, while the other lives quietly on earth and when the traveler returns he realizes that his brother is older and for him only a few years have passed. If time were not elastic, this maneuver would be impossible, the speed of light would not be the maximum attainable limit, and a space-time perforation would be easily produced.

The elasticity constant of the space-time mesh has not yet been properly researched, however, we can already glimpse the importance of this characteristic of this entity that has been shown to be of such importance. Not even a photon of energy would move if space-time were not elastic, the conditions for the existence of anything require a balance maintained by elastic potentials and constants, characteristics and often unique characteristics of each thing.

Hence the questions, what is the elastic limit of space? What happens after exceeding this limit? How can we make better use of this aspect? For the next works of a more restricted nature to the subject, there will be details on these issues that are beyond the scope of this proposal.

REFERENCES

DE SOUSA, A.A.; DE FARIAS, R. F.; Chemical kinetics: theory and practice. Atom, 2nd ed. Campinas- SP, 2013.

REIS, M.C.; BASSI, A.B.M.S.; The second law of thermodynamics. Química Nova, Vol. 35, No. 5, 1057-1061, 2012.

MUTZENBER, L. A.; VEIT, E. A.; Silveira, F. L.da; Elasticity, plasticity, hysteresis... and waves. Brazilian Journal of Physics Teaching, v. 26, n. 4, p. 307 - 313, 2004.

BORES, A. P.; SCHMIDT, R. J.; Advanced mechanics of materials. Copyright Materials, United States of America, 2003.

CALLISTER,W.D.Jr.; Materials science and engineering an introduction. John Wiley & Sons, Inc., New York, 1991.

PAVIA, D. L.; LAMPMAN, G. M.; KRIZ, G. S.; VYVYAN, J. R.; Introduction to spectroscopy. Translation of the 4th ed. Americana, Cengage Learning, São Paulo, 2010.

DALCIN, G.B.; Materials testing. URI – Santo Angelo, Industrial Mechanical Engineering report, Santo Angelo, 2007.

MASCIA, Nilson Tadeu; Traction, compression and hooke's law. Department of Structures. Workbook of the course of the Faculty of Civil Engineering, Architecture and Urbanism, State University of Campinas-UNICAMP, São Paulo, 2006.

First view of general theories of elasticity

Theory is a model of maximum explanatory reproducibility about a given phenomenon or set of them. Although it should be accepted, it should never be considered finished and unquestionable.

Every thing, material or not, as seen in chapter 1, has peculiarities that influence its mechanical and/or chemical behavior from the point of view of elasticity. Thus, theories or models have been developed to explain the elastic behavior in each group and type.

From now on, we will focus on materials and no longer on the entities highlighted above, after all, our first idea and foundation of elasticity arises from materials, that is, from that which is solid, has form and preserves its integrity within its own resistive limit, determined by the chain bonds of its atoms.

Today, about five main classes of materials are accepted, these being elastic, elastoplastic, hyperelastic, viscoelastic and superelastic, which have several theories that aim to explain them, among which we highlight those of greatest interest for an introduction to the elasticity and behavior of solids.

Materials are divided into two main categories, linearly elastic and nonlinear, so for those that faithfully obey Hooke's law, it is considered as linear elastic solid, and the others whose strain stress graph is nonlinear, several other models are used to explain them.

Among the main elastic theories of nonlinear solids, the Von Mises/Tresca and Drucker-Prager theories that describe the behavior of elastoplastics, the Mooney-Rivlin/Ogden and Blatz-Ko theories for hyperelastic materials, Maxwell's and Voigt's theories for viscoelastic materials, and Yong's theory for simple elastic materials stand out. It also has the Nitinol model for describing superelastics, such as alloys with reforming memory.

In addition to the type of material, its geometric shape also significantly influences its elastic behavior, for this reason, in material testing, the modulus of elasticity of the body considers several factors specifically to accepted normalizations and not simply by stress/strain.

From this perspective, it is necessary to observe some concepts that disrespect the characteristics of some materials. The first concept is that of isotropy, a characteristic that gives the material asymmetry of elastic modulus in all directions of the body, therefore, an isotropic solid (from the Greek *ἴσος* = equal, *τρόπος* = action of itself returning to something), in theory, does not assume different moduli of elasticity for other directions of the axis or plane of request.

Anisotropic solids are opposite to isotropic solids, presenting different elastic moduli for different positions of the stress axis. When these differences correspond only to differences in elastic modulus in perpendicular directions to each other, but still present a certain symmetry for the other directions or directions, the solid is called orthotropic, which is a designation for an anisotropic solid, but with a certain order and specificity.

35

Equations 9 and 10 show respectively the coefficients of anisotropy (r) and isotropy of a material. The calculation of these characteristics is of crucial importance for choosing the applicability of the material, its shape and directionalities of the maximum and minimum elastic strength, among other possible ones, such as resistance to vibrations and diversity of requests (direction, direction and speed).

In equation 9, it is the deformation of the width and it is the deformation of the thickness, however as there is a certain variation due to the error of the measurement, we will calculate the natural logarithms as rations, obtaining a value that can even be considered equal due to the high degree of approximation. Thus, we set as the numerator the ln of the initial width over the final and as denominator the ln of the ratio of the product of the final length to the final width over the initial length to the initial width.$d_l d_e$

$$r = \frac{d_l}{d_e} \cong ln\left(\frac{L_i}{L_f}\right) / ln\left(\frac{C_f.L_f}{C_i.L_i}\right) \qquad \text{(Eq. 9)}$$

There are also planar and normal anisotropic coefficients when considering anisotropy from the point of view of the plane of a sheet, for which the calculation formulas used are usually those exposed in equations 9.1 and 9.2 respectively, for sheet metal. Remember that when it comes to materials, some equations can be modified to explain characteristics of other materials and satisfy the conditions of the experiment.

$$r_p = \frac{r_0.2.r_{45}.r_{90}}{4} \qquad \text{(Eq. 9.1)}$$

r_0, r_{45}, r_{90} are the coefficients measured from different angles of the laminar region. Whereas, when and the material is isotropic. See that this is the difference that r of the material presents in the plan. $r_p \neq 1 \Delta r \neq 0 \Delta r$

$$\Delta r = \frac{r_0 - 2.r_{45} + r_{90}}{2} \qquad \text{(Eq. 9.2)}$$

As seen, it is possible to describe an isotropic system apart from the calculation of an anisotropic system, however, when the class of the material has an isotropic nature like most steels, for example, we can through the equation k, discover its isotropic character, using the Poison coefficient (n), where we see the relationship of Young's moduli (Y), shear/shear (C_s) and compressibility (C_p). Being the transverse deformation and it is the longitudinal deformation. $d_t d_l$

$$n = -\frac{d_t}{d_l} = \frac{3C_p - 2C_s}{6C_p + 2C_s} = \frac{Y}{2C_s} - 1 \qquad \text{(Eq. 10)}$$

Note that there is a tendency of n < 0 in some materials and, when n is between -1 and 0.5 it can be said that the material is isotropic, when these conditions are not met, the material is anisotropic.

Now, we can start our study on the properties of linear and nonlinear elastic materials. Knowing that both can be anisotropic and isotropic, belonging to either class, at least in a given request interval.

Linearly elastic solids

These materials are mainly represented by metals and their alloys, so linear solids have Robert Hooke and Young's theory as their main explanatory theory. From

37

which all other theories were created, including for nonlinear solids.

Hooke postulated that there were proportional relationships between the force applied to a body and its deformation, however, he did not delve into the subject, most likely because of his rivalry with Newton. In 1802, the subject was taken up again by Thomas Young who discovered that there was a constant between force and strain so that they were in fact proportional, although he expressed this postulate in such an enigmatic way that it was not well understood in his time, it became together with Hooke, although with less splendor, the basis of the elastic theory.

Much of the knowledge about elasticity, especially the concepts and understandings about tension and traction, were only clarified from the theories of Isac Newton and Augustin Louis Cauchy, who from then on, gained increasingly complex equations, given by scientists encouraged by the apogee of the industrial revolution and post-modernity.

Today, for example, it is understood that the elastic constant is originated as a function of the properties of the chemical and physical bonds between atoms, the nature of the atoms, the atomic dispositions and arrangements in relation to the geometry of the solid body and its position of request, that is, if the body is anisotropic, if it is isotropic, the modulus will be the same regardless of the plane of the request.

It is important to remember that elastic solids only have linear behavior up to a certain point of the required

1st Chapter

load demand, from then on the deformation can be plastic, thus, they are only linearly elastic for small loads.

Hooke's law is applied without modification to linear and isotropic solids, although currently its generalization is used, in which the discrepancy is usually discounted from the elastic constant or mathematical relations are used to reach conventional values. Therefore, in the course of this topic we will gently address non-classical theories of linear elastic behavior in a complementary way.

Initially we can use as a starting point for compression of elastic bodies, many of which we use for springs, since some definitions of springs apply to a wide range of materials with different shapes, for less superficial mathematical details of linear elasticity see topic springs in chapter 4.

We know that for a one-dimensional element we admit the law F = k.x, however, for isotropic three-dimensional we generalize the previous equation to equation 11, in which we can describe a product between the matrix (M) and the deformation (x), this matrix being formed by a set of coefficients that, in short, can be described as a function of only three, Poison's coefficient (C_p), Young's modulus (Y) and shear coefficient (Cs), in this way we can write equation 12.

$$F = M. x \qquad (Eq.11)$$

$$C_S = \frac{Y}{2.(1+C_p)} \qquad (Eq.12)$$

Note that at this point, we only interrelate coefficients to obtain the cut-off coefficient, so when we want to make an analysis of the stresses acting on an elastic body, we must enter into a deeper study of Newton's concepts and equations, which is not the purpose of this book. Therefore, simply, we can describe the behavior of stresses from two perspectives: Normal stress and shear or shear stress.

Equation 12 is the simplest way to represent tension () in a body, assuming isotropy, we can write that the ratio between force (f) and area (A), which at other times describes pressure (for fluids), here describes tension.τ

Normal stresses can be classified as two types, compression tensile and tensile stress. Which act perpendicularly on the part and are differentiated because for traction it is understood that >0 and for compression <0, therefore f>0 and f<0 respectively.$\tau\tau$

$$\tau = \frac{f}{A} \qquad \text{(Eq.12)}$$

The shear stresses () of the cross-section can be further described, using equation 13, where τ_cQ is the static moment, V is the internal shear force that results from the stress, I is the moment of inertia and l is the width of the cross-section. These stresses are important to define the safety elastic limit and maximum elastic limit.

$$\tau_c = \frac{QV}{Il} \qquad \text{(Eq.13)}$$

From the understanding of stress, we can calculate the energy spent on a strain by calculating the work done, so by integrating the strain interval (x) into Hooke's law we arrive at equation 14. where is the variation of the elastic potential energy.Δe_p

$$W = \int_{x_1}^{x_2} k \, . \, xdx = \frac{k.x_2^2}{2} - \frac{k.x_1^2}{2} = \Delta e_p \qquad \text{(Eq.14)}$$

There are many engineering systems and equations that are necessary to characterize a linear material much more complex than the one demonstrated above, however, this is the basis and enough for an initial view, since linearly elastic solids are the simplest.

Nonlinear Elastic Solids

Nonlinear solids are those that do not have a line on the graph of the force by strain function, but a parabola or other more complicated curves. These solids are much more numerous than linear solids, since they represent almost everything that is known, and can also be divided into categories.

Generally, nonlinear solids have a non-crystalline molecular structure and covalent bonds, and may have many intermolecular forces acting on their constitution. For these bodies, the hypothesis of physical linearity is not mechanically adopted, only of geometric linearity. Thus, some of the principles of linear elasticity can also be adopted for nonlinear elasticity, such as displacement gradient, tensor field, and boundary conditions.

However, we can write in a generalized way the equations of nonlinear elasticity using the Cauchy model that relates the tensor of the applied stresses with the tensor of the deformations of a body, by means of equation $14, \sigma_{ij}\varepsilon_{kl}$

$$\sigma_{ij} = f_{ij}.(\varepsilon_{kl}) \qquad \text{(Eq.14)}$$

Being response functions of the material. It is worth remembering that a more rigorous treatment mathematically with deductions from the expressions exposed here is beyond the scope of this work, since for this it would be necessary to have at least an introductory knowledge of tensor analysis. But the tensor terms found in equation 15 can be expressed by matrices, a more well-known mathematical entity. f_{ij}

Analytically, Cauchy's model consists, in the case of an isotropic material, of

$$\sigma_{ij} = \phi_0\delta_{ij} + \phi_1\varepsilon_{ij} + \phi_2\varepsilon_{im}\varepsilon_{mj}, \qquad \text{(Eq.15)}$$

In this tensor equation the convention of Einstein's sum applies, where repeated indices mean sum in the three dimensions of space. The symbol is the Kronecker delta tensor which is zero if and one if, thus being represented matrix by the identity matrix: $\delta_{ij} i \neq j i = j$

$$\delta_{ij} = (1\,0\,0\,0\,1\,0\,0\,0\,1) \qquad \text{(Eq.16)}$$

In addition, the coefficients , and are functions of the invariant contractions of the strain tensor. The tensor representation is capable of synthesizing in a simple notation the stresses and deformations in all directions of space, and can even be generalized to larger spaces. $\phi_0\phi_1\phi_2$

It is interesting to note that in the first-order Cauchy elastic model for an isotropic material and bodies that have nonlinear elasticity can also be represented by the same parameters of Hooke's law, considering the initial conditions of zero stresses and strains, the generalized Hooke's law is reduced, that is,

$$\sigma_{ij} = \lambda \varepsilon_{kk}\,\delta_{ij} + 2\mu\varepsilon_{ij}, \quad \text{(Eq.17)}$$

Being and Lame's coefficients. While it may be called a stiffness modulus, it is not given a particular name. Mathematically, it is the trace of the tensor of the deformations or the sum of the elements of the principal diagonal of the matrix that represents the tensor $.\lambda\mu\mu\lambda\varepsilon_{kk}\varepsilon_{ij}$

Models of a higher order than the first can be found, but evidently the values of the functions , and must be different from those obtained previously.$\phi_0\phi_1\phi_2$

This is because, unlike linear solids, nonlinear materials can respond to load demands with varying parameters for each order, which is why there is a curve in the diagram, as the constituent forces of the material react differently for each stress value.

As for work and elastic potential energy, in general terms the same equation can be used as linear elastic systems, since they depend on force and displacement, not on the way it occurs.

Elastoplastics

The elastoplastic model is used for ductile solids at room temperature, such as most metals and some polymers, in which they are considered as determinants of behavior, yield stress, limit value of shear stress reached – Tresca's criterion – and the flow given as a function of a state of stresses is found on the surface of the material – Von Mises criterion –.≥ 0

There are many criteria capable of characterizing a material as part of the elastoplastic model. However, the former summarize its basic properties and allow more advanced models to be applied.

Its study occurs due to a notorious uniqueness of some materials in behaving both elastically and plastically. In general, most plastic materials can be classified as elastoplastic, as long as they also present elastic behavior, however, the way in which the transition between the elasticity limit and yield strength occurs is differentiated, especially for those that fit better only in the plastic model, which tends to yield and hardening characteristics.

For the elastoplastic model, the deformations that often seem elastic are not completely elastic, but are also plastic, that is, permanent. The way the material starts to respond to the load demand within an elastic regime in the plastic phase changes constantly, and this is one of its main characteristics.

This change is determined by the law of hardening, which results from the fact that, in some cases, the hardening of the material occurs during the plastic regime, which when it does not occur, it is said that the body is perfectly plastic. These conditions are in which there is no change in the boundary and surface of flow.

44

The hardening of elastoplastic materials occurs in basically three types, which may or may not lead to breakage. These are:

Isotropic hardening – when there is an increase in the flow surface, but not in position and shape;

Kinematic hardening – The flow surface does not change shape, but changes position;

Mixed hardening – The flow surface moves and increases in size.

The latter, in a real loading, is what finally produces, after hardening, the last phase of the plastic regime, the restriction. Which is the point at which the thinning or "neck formation" of the body occurs at a point, usually in the center – depending on the test – leading to separation or shear. This can be represented in graph 3.

Graph 3. Graphic form for an elastoplastic regime.

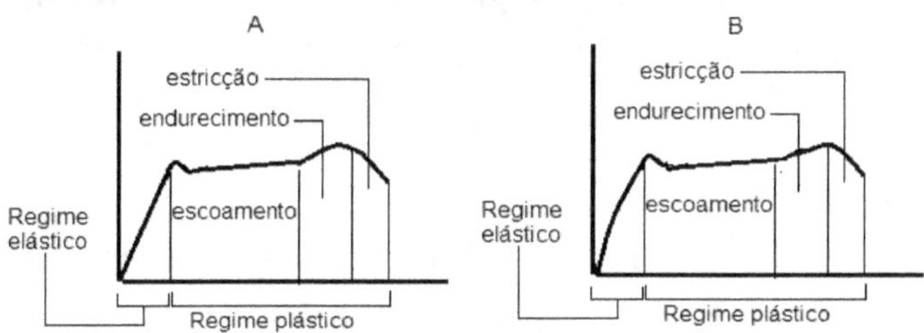

Source: designed by the author himself

Kinematic hardening is the way that load cycles modify the elastic regime when the plastic phase is reached. However, the understanding of this phenomenon is beyond the purpose of this book.

The graphs show behavioral bodies, in (A) linearly elastic and in (B) a nonlinearly elastic, bodies with these characteristics are best represented as being elastoplastic. Note that the behavior of the plastic regime is shown with strong evidence, along with the elastic, the behavior is graphically characterized in a more general way by graphs 5.

Graphs 5. Some types of mechanical behavior.

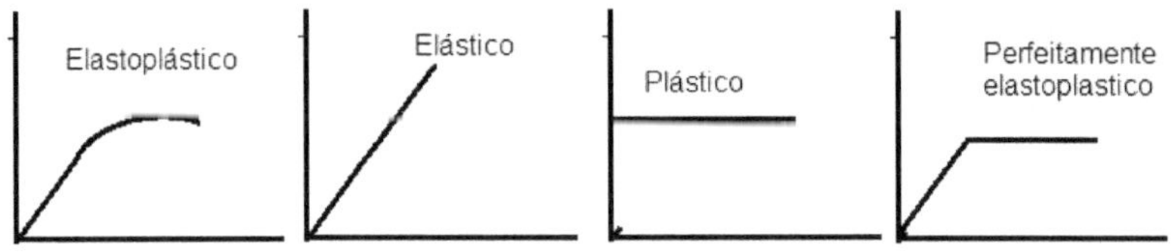

Source: designed by the author himself

The interpretation of these models (elastic and plastic) when occurring in a single material, provides and justifies the existence of the elastoplastic model. Since even in a generic way it has to be admitted that the behavior of materials is so different, one in relation to the other, that most of the time fused concepts are used to explain their essence and properties.

Hyperelastic

The hyperelastic solids model is applicable to materials that "completely" restore the energy used to stretch them. In which, for Cauchy, the state of the stresses does not depend on the path by which the deformation occurred when a force is employed, only on the final and initial energy state.

Elastomer polymers are good examples of materials with great nonlinear elastic capacity, they are good

46

examples of hyperelastics, implemented especially after the modern age, being called hyperelastics since the nineteenth century.

Elastomers generally have a Young's modulus of around 0.1 Mpa and a distortion modulus of around 10 Mpa, which can be even smaller depending on their nature and synthesis technology.

There are many models that seek to explain these materials, especially because they present certain peculiarities within their long interval of response to requests.

Hyperelastic solids conserve the energy expended in the effort required to deform them, restoring it almost completely. Thus, an elastomer can be considered as a thermodynamic system and its demands calculated in general form of energy. As shown in the thermodynamic equation of rubber in the topic "polymers" of chapter 3.

Graph 5. Stress deformation behavior of a hyperelastic.

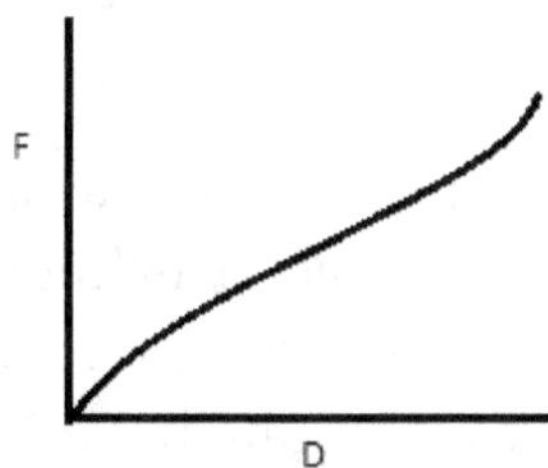

Source: adapted by the author himself

According to Cauchy's work, the best relationship to describe hyperelastic drugs is not force and strain as in Hooke's systems, because the best characteristic of this system is the energy response to the load demand. Thus, it is possible to describe the responses of the materials as

47

a nonlinear response of strain and energy, especially because most hyperelastic cells are nonlinear, and can be represented in a general way by graph 5.

Generally, hyperelastic systems, whether linear or not, have almost total energy restitution, the relationship between applied force (F) and deformation suffered (D) is represented, almost always according to graph 5. Especially for non-linear.

These solids are materials that most of the time do not have crystallinity and the organization of the molecules is random, connected by long and coiled chains. Whose cohesion is given by intermolecular forces that keep a chain coiled over itself and linked to others.

Viscoelastic

This model is applicable to materials that have viscous and elastic behavior simultaneously. It is generally considered that part of the load is elastically cushioned, that is, the energy is dissipated. And another fraction of the body reacts elastically, returning the energy in the effort.

As you can imagine, there are viscoelastic models that comprise solids whose elastic character is linear and nonlinear. The shape of the graph depends on the variables and the viscoelastic character, and it can also be viscohyperelastic, for materials with a hyperelastic regime, but which also has viscous behavior.

Elastic behavior modeling

As seen, for the damping solids we can use Newton's viscoelastic model, for the association of a plastic material with an elastic (itself) we can use the elastoplastic model of Prandtl-Reuss.

For the association of an elastic and a viscous model we use Maxwell's viscoelasticity, for parallel association of an elastic with a viscous we use Kelvin-Voigt's viscoelasticity, for the junction of the previous model with an elastic we use the viscoelasticity of three "Boltzmann" models, for a parallel association of a plastic with a viscous model we use Bingham's viscoplastic model and for the association of an elastic model to a viscoplastic in series we use the Hohennemser-Prager theory.

There are many models to explain the elastic nature of materials, which continue to emerge. For more details, the bibliography of the chapter should be consulted, as it only aims to introduce such contents.

REFERENCES

BALL, J.M.; Some open problems in elasticity. In Geometry, Mechanics, and Dynamics, Springer, New York, 2002.

BISCHOFF, J.; Viscoelasticity Constitutive response of tissue at large strain rates. Esci 274 Mechanics of Biomaterials, The University of Auckland, 2002.

SANDOVAL, C.F.B.; Elasto-plastic models and their influence on the design process of mechanical components. University of Brasilia, Master's thesis in materials sciences, Brasilia, 2014.

M 'UTZENBERG, L. A.; VEIT, E. A.; SILVEIRA, F.L. da; Elasticity, plasticity, hysteresis... and waves. Brazilian Journal of Physics Teaching, v. 26, n. 4, p. 307 - 313, 2004.

PASCON, J.P.; CODA H.B.; Constitutive models for hyperelastic materials: computational study and implementation. *Cadernos de Engenharia de Estruturas*, São Carlos, v. 11, n. 50, p. 131-153, 2009.

PIMENTA, P.M.; Fundamentals of Solid and Structural Mechanics. Handout, Polytechnic School of the University of São Paulo, São Paulo, 2006.

Principles and Laws of Simple Elastic Bodies

"All bodies, although distinct, have characteristics in common. Which, when ordered and understood, can be classified as resulting from a principle or law for the bodies that comprise a given category."

A simple elastic body is defined as a body that does not establish relations with a complex system, so a simple elastic body is one that under the action of a force does not depend on another to influence its deformation. For example, an iron bar, a concrete bar or even a brick or rubber is mentioned. All of these can be considered units for the formation of a complex elastic system, if properly combined and organized.

For each of the bodies mentioned there are rules and, sometimes, very specific theories that can be easily understood when one resumes the genesis of elasticity and its thermodynamic reasons.

Note that the reason why elasticity exists is the establishment of an equilibrium energy/matter conformation, which when passed a disturbance by an external force tends to restore the original conformation. We will see each of these factors separately in general to understand what the causes and laws that govern elastic behavior for each situation really are through a brief review.

Let us consider the most common solid materials in terms of elasticity to understand the properties that can interfere with the elastic characteristics of materials. In general, six properties are of relevant interest in the characterization of materials, being mechanical, electrical, magnetic, thermal, optical and deteriorative as groups of these properties since they can be divided into several subclasses.

Remember that most existing materials can be classified into one of the following categories: ceramics, metals, polymers, composites, crystals, biomaterials, semi- and superconductors — there are other more modern classes that will not be addressed due to the interest of the book — for which elastic behavior has its specificities, although it always seems to be related to the second law of thermodynamics.

Therefore, it is necessary to understand that different materials have different properties and the way they intersect is unique and corroborate the existence of elasticity in existing bands of very specific Yang modulus. At this point, the existential range is considered to be the range that goes from the minimum to the maximum modulus of elasticity for a given type of material or class.

All the properties of materials are interconnected, so when one of them is subjected to stress or energy demand, the whole body feels, therefore, all the other properties, although some more than others, depending on the stimulus. For example, when a body to 273 K is

52

heated to 300 K its elastic properties change as a function of this increase in temperature, in general the body becomes more flexible and more rigid with the opposite.

The interatomic displacements or vibrations introduced by the increase in temperature make the body more and more ductile, increasing the ease of sliding of one atom over the other, however, the forces that kill them together, and therefore the elasticity of the body, decreased.

There is a temperature sweet spot before elasticity is transformed into viscosity. Thus, if a metal body heats up too much, it can cancel out other mechanical, thermal, etc. properties even before the phase change. Which occurs as a function of the Newtonian law of action and reaction, thus it can be postulated that:

> *Every stimulus of a property in a body alters the others that are related to it.*

The above postulate expresses that if a body is forced to increase its hardness by intervention via temperature, pressure or other condition. It is very likely that all other properties will be modified, since the mechanical properties of materials are directly related to their structure and chemical composition, therefore, modifying a parameter of structure or bond strength physically or chemically, falls on the other properties of the material.

Therefore, modifying material properties requires a thorough study of the consequences that these

53

modifications will bring on the other characteristics and properties of the material, since they can satisfy a utility parameter and be unfavorable to use due to a new characteristic that is unfeasible for the desired application.

We will make a brief analysis of the properties and elastic behavior in some of the most common materials to understand how the above postulate is related to these classes of materials.

THE PROPERTIES OF MATERIALS

Materials formed by ordinary matter generally have a definite and proper structure, as shown in chapter 1 that we will discuss in more detail below. These structures, combined with the composition and transit of energy in the system, define its properties. Which are:

Elasticity: it is the ability to restore its original size by removing a deformation effort;

Resilience: is the ability to absorb and release energy through elastic behavior;

Ductility: represents the capacity for deformation before breaking, that is, the ability of atoms to move over each other by applying a force without necessarily breaking the material.

Plasticity: It is the inability to maintain the reticular structure or molecular arrangements of the material through the application of effort. It occurs after the elasticity limit and unlike ductility, it causes structural modifications, representing a behavior that leads to the shear (breaking) of the material. The

54

plasticity limit is measured at the point of safety prior to the start of the elastic regime and the energy required for deformation usually decays until failure occurs from this point.

Frailty: The ability to lose its original shape inelastically and without satisfactory ductile response upon exertion, usually the physical response given is fracture, shattering, and sometimes degeneration. Brittleness can be measured by calculating the maximum stress that the object can undergo before breakage. Materials that suffer fracture due to little deformation are considered fragile.

Toughness: It is the resistance to impacts/shocks before breaking, mathematically it is the total area of the stress-strain diagram. The elastic, ductile and plastic responses of the material before shear or fracture.

Hardness: It is the degree of resistance to the effort inferred to tear off pieces when rubbed against another, that is, resistance to grooves.

Electrical and thermal conductivity: it is the ability to conduct heat and electricity, it is related to the ability of electrons to propagate in the body forming current (in the case of electricity) and of atoms to vibrate at infrared chain frequencies, which causes heat transport. However, a material can be a good conductor of heat and not electricity, vice versa, or both, which depends on the nature of the material.

Thermostructure: is the ability to modify its reticular or molecular structure before undergoing a change of state as a function of temperature. Which, when existing in a material, can change its properties in a non-gradual way, more drastic changes, which is common in some metal alloys, which can generally have two or three structures, and may have more.

Hardenability: susceptibility to rapid cooling, causing as a response not to rupture due to thermal shock, but to a reduction in elasticity, ductility, toughness and increased hardness and tensile strength, a characteristic generally found in steels, which when going through a tempering process, modify their reticle permanently. It also causes the emergence of internal tensions.

Corrosion resistance: the ability of a material to resist the action of external oxidants, forming areas of high stresses capable of causing cracks that propagate through the material.

In addition to these basic properties, depending on the material, others may arise, as well as be specific to a particular group, class or type. These properties can be of a mechanical, chemical, biological or physical nature, in general man makes them a parameter for control and analysis.

For example, photolithicity is mentioned, which we can use to describe the susceptibility of materials to photons in the most varied spectral regions, a property of some polymers and ceramics for military and industrial use (use still under study), since some are very fragile to certain frequencies of light, although they can be resistant to great mechanical stresses.

A property that is still unconventional, but tends to become popular in engineering, especially in the polymer industry for roofs and paints.

Other more common properties, such as color and brightness, are also the subject of exhaustive studies and can significantly influence other properties, since they result from intrinsic characteristics of the materials, all of which are interconnected.

HOW MATERIALS ARE ORGANIZED

Materials are divided into three general classes regarding their structure and chemical bonds: metals and their alloys, ceramics and polymers, also considering composites that are (very specific) mixtures of classes or types of materials, so they are not elementary, more groupings and can vary greatly.

Scheme 1. Organization of materials.

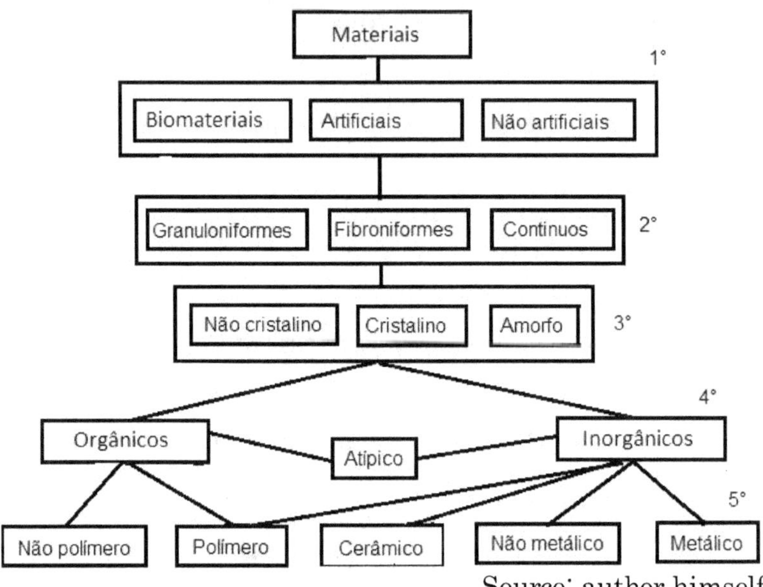

Source: author himself

These classes have subclasses and groups, which can be further divided and interconnected by molecular, responsive and original characteristics. At this point, a way to organize the materials is proposed, since the great disagreement in carrying out a general sequencing of all materials in an orderly way. What is didactically proposed in schema 1. Where it is shown how we can organize them according to origin (1°), microstructure (2°), molecular structure and chemical bond (3°), nature (4°), class (5°).

Atypical materials are solids that do not fit into any of the other origins and types, are rare and usually found in research and frontier knowledge.

Scheme 2. Materials formed by combinations.

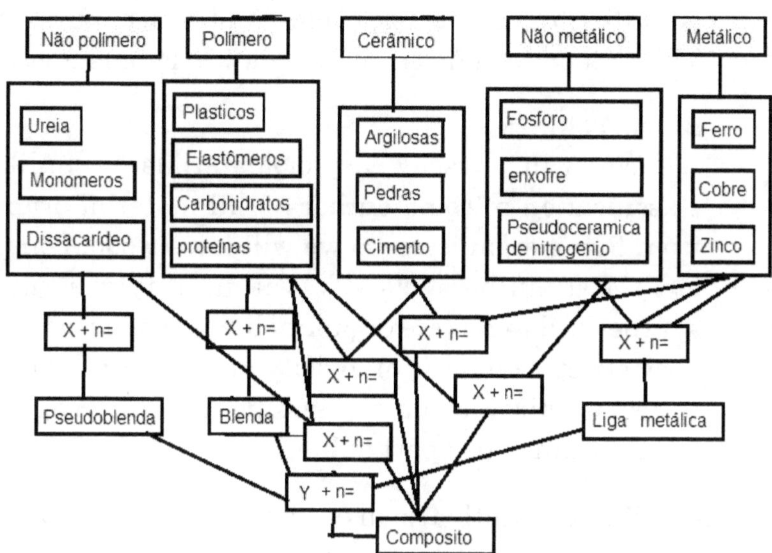

Source: author's own example

These classes and subclasses (the latter was omitted in scheme 1, e.g. elastomers, glass, plastics) can still be combined, generating materials with new properties, some of these "mixtures" are very important and were also classified according to the original products and the characteristics of the final materials, where scheme 2 is generally adopted.

Note that only a few subclasses of materials were presented, however the classes of combinations are complete. It is necessary for a combination x + n materials or even Y + n in the case of composites, which can also be recombinations.

The elastic behavior of materials sometimes varies according to their class, origin, and molecular structure. This can be seen in the above below for some materials.

BIOMATERIALS

We will classify as biomaterials – unlike the classification of some authors – all material originating from living organisms, so we will have wood, bones, silk, etc. Most authors classify within this the materials synthesized to replace organs of living beings, which must obey strict standards of toxicity and resistivity, however, it seems fairer to us to classify them as bioproteoids, which means false part of life.

We constantly use in our lives the remains or parts of other dead and living beings to build things and take advantage of them. Elephant ivory, cattle bones and horns, wood, leather, and shells are good examples of this; However, understanding the behavior of elasticity in each of them requires substantial knowledge, but in general we can state that:

> *The elastic behavior in these depends among other things on their state of conservation, the amount of water, the organizational structure of their fibers or particles and their composition.*

Biomaterials are diverse and can be divided according to their origin, class, type and structure, they are the most difficult to classify in terms of their properties, since they are diffuse, and can be from polymers to ceramics. However, observing the principle highlighted above and considering that they can be found in two main natures:

Organic: those that are presented in the form of organic compounds or their structure is formed by conglomerates or structures of organic material;

Inorganic: when formed by inorganic structures or conglomerates;

So that they are subdivided into types according to their origin and structure, thus, we can say that the wood of the cassava stem is of a different type from the wood of its roots. We say that the material has the same origin, biomaterial obtained from a plant, however, it is of a different type, because the place of extraction and structure is different. Since the origin has to do with the being that provided the biomaterial, fungus, bacteria, vegetable, animal or protozoan.

If a biomaterial is obtained from the horn of the ox it may be of the same type, origin and class as a second obtained from the horn of a buffalo, but it will not be of the same type.

Biomaterials are generally similar to each other, although they can sometimes be divergent. For example, the most common biomaterial of plant origin is wood, which is quite different from bones, except for the fibrous and tissue-like aspect they present, which in a way characterizes them, since most biomaterials are tissue-shaped, that is, they have an organized formation,

61

arrangement and structure with a predominantly fibrous structure from cellular arrangements or that, due to their synthesis, they resemble them. However, when comparing bones to a pearl, the structural difference is eminent.

The elastic behavior of biomaterials is directly related to such characteristics, for example, woods (and other materials) are very dependent on the amount of water inside them to correspond to a certain mechanical performance. Whose elastic modulus depends on the concentration of moisture – whether it will be beneficial or not – depending on how the water combines with the fibrous and molecular structures of the tissue. Therefore, in each biomaterial there are optimal points of water/elastic modulus concentration, which is crucial for the best use.

For biomaterials in general, different elastic moduli are found for different parts of the same body. The elasticity in biomaterials, such as in latex – which is also a polymer – of the rubber tree and the chitin of the brittle shell of insects are so diverse, that the only thing that resembles them is the fact that both are biomaterials, that is, they were produced by living cells and as such leave "manufacturer's impressions", they are an important reminder that life meets its demands with what is available in the environment and builds its own materials generating new biochemical demands when solving a problem.

62

CRYSTALS

Crystals are bodies defined as such as the organizational structure of the atoms of their molecule(s). In them, atoms are in the form of cells and have ordered structures, while in amorphous materials this is not the case. The word crystal came from Greek to designate ice, initially used for quartz crystals. Today we can define them concisely as:

"A crystal is any solid endowed with a crystalline system and wild cells, a homogeneous, anisotropic solid in which at least one of its vector physical properties is discontinuous with the regular internal order of its atoms or ions. "

The crystals have internal arrangements classified into 14 Bravais cells and 7 crystalline systems, as shown in figure 1 (chapter 1). Such structures give special characteristics to these materials. Although there are occasions when the characteristics can be so similar as to be mimetic, each crystal or crystalline body is unique.

Crystals usually have large modulus of elasticity, so their resilience is small, especially for mineral crystals, they are translucent or not and have different resistances to impact intensities. The elastic behavior of crystals is little explored, however, metalloids and metal alloys have crystalline arrangements and are more explored for their elastic properties, which will be better explored in the appropriate topic.

Crystals can form complex structures with quite specific chemical and physical characteristics, not much

63

is known about the dominance of the physical properties of crystals. Therefore, it is not possible to say how elastic it is only by its crystalline structure – which is only measured – but by the combination of chemical composition with physical characteristics.

Ceramics are probably the most impactful solids when it comes to crystals, especially because they have some of the most desirable specimens, such as most precious stones. The elastic behavior of these, when they do have a crystalline reticulum, is influenced by temperature, although not significantly. Other ceramics, such as those derived from clays, can have their elastic modulus modified with the increase in temperature, in general, increasing their mechanical resistance.

Some compounds that have a crystalline lattice can also vary their structure to conformational arrangements more favorable to energy flow demands, thus, a compound with cubic unit cell x can modify its cell to y from a certain temperature, being able to vary n times, this is because most materials suffer expansion with temperature (degree of agitation of the molecules) so that in certain demands, A given structure becomes more stable/favorable than another to the new distance between the atoms and their frequency of vibration.

It is based on this property that some solids change their color when heated, either because their electrons (the ones that produce the visible spectrum) interact differently or because there is more or less space between the atoms giving passage to light. Whether the color

64

change will affect the modulus of elasticity will depend on the cause and the degree of change in this structure, because as already mentioned, all properties interact with each other, as they depend on the structure and chemical bonds.

BRASS

Metals are chemical elements characterized by having a characteristic brightness and color, the ability to conduct heat and electric current; Among other properties that make them special, at the molecular level we can highlight the way their electrons interact when atoms in a metallic body, behaving like a fluid that covers the entire surface of the metal, so the theory that explains them is called the Electron Sea Theory.

The molecular structure of metals follows the rule of crystalline cells, however, they are monophasic or biphasic and the sliding of atoms is more efficient than in most salts and crystalline molecules, because they are impregnated with a sea of electrons. Contrary to what one might think, the corporeal dimensions of the metal are not stopped, but they can form clouds that easily extrapolate this boundary with more intensity than the electrons of the body dimensions of other substances – except superconductors ⁻.

Scheme 3. General classification of metals.

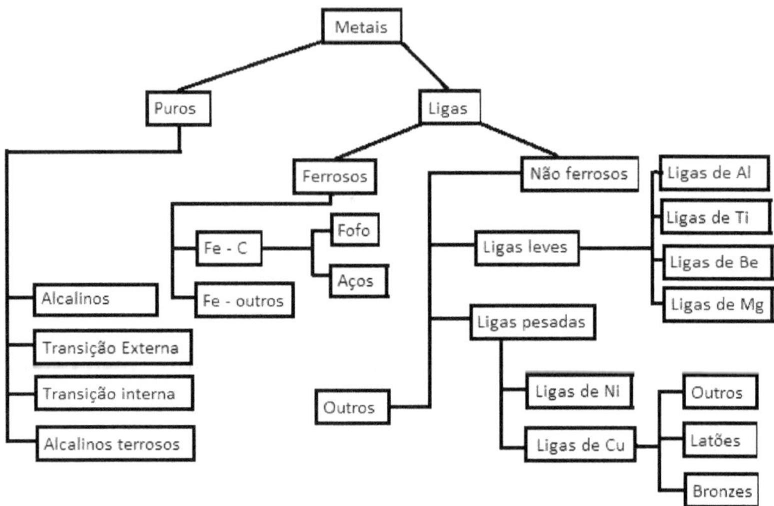

Source: author himself

Due to the electronic behavior, metals can present marked toughness, hardness and ductility that considerably influence their modulus of elasticity. This varies a lot from one element to another and sometimes forces engineers to make metallic or mixed alloys with other non-metallic and semi-metallic elements in order to obtain the characteristics demanded by the industry. It is worth remembering here that even metal alloys dry the general rules of the metallic elements, with slight but important divergent properties, so metals are classified according to diagram 3.

The number of metallic compounds, especially alloys, including polymers, is increasing, so that soon, it is possible that a new division of compounds classified as "other" into a group of superelastic and ultralight compounds will occur.

Metals can have their modulus of elasticity changed in a reversible way, with the addition or withdrawal of energy, as shown in the previous topic, due to their thermostructure. As well as small changes in temperature and electric current transit – on special occasions for the latter – which is also valid for alloys.

In this perspective, the modulus of elasticity can vary as a function of the environmental conditions and energy composition of the material, with temperature values for elastic modulus varying from element to element, even without its structure changing significantly, only with simple expansion. Therefore, it can be said that:

"The elastic behavior of metals depends on their structure, temperature, electrons in transit and chemical composition."

Other materials as well as metals are also influenced by temperature on their elastic capacity, but the sensitivity of metals to temperature is responsible for the most pronounced change in their elastic behavior. This is especially considered in the manufacture of parts for internal combustion engines, after all, if not considered, can lead to false load capacity measurements.

This phenomenon can be explained by observing that the increase in temperature consists of the increase in *inter* and *intramolecular* vibrations, so that atoms move more easily. Making the higher the temperature, the lower the modulus of elasticity (it is not a rule, there are

67

exceptions, also remember thermostructure), so the material becomes more and more flexible the higher its temperature until it changes phase, when it acquires viscoelastic behavior.

Currently an (atypical) metallic alloy of gold and polyurethane is the most elastic known, created by the University of Michigan – USA in 2013, it can be stretched up to 5.8 times its size, maintaining metallic properties, such as electrical conduction around 35 S/cm, maintains some characteristics of a metal alloy, on the other hand steel, iron carbon metal lyric has an elastic modulus of 200 Gpa – the modulus of elasticity of common glass is around 90 GPA – that is, it is not very flexible, and could never be stretched to these dimensions.

Although steel is less elastic than glass, it has greater ductility, so it can be bent more easily and that is also why they are less brittle. Many alloys are produced in order to increase resistance to corrosion, impacts and increase in elastic modulus, which classifies them as still belonging to the class of metals, are their bonds and chemical structures.

The low elasticity of metals – which thanks to technological advances is no longer the rule – sometimes leads to their replacement by other materials, for example, carbon fiber that replaces steel in helicopter vanes, which in addition to being very resistant admits greater elasticity. However, superalloys based on Niobium, an element whose largest deposits are found in

Brazil, are used to make highly resistant turbines for rockets.

In general, it is desired that the materials have good elasticity, being able to resist impacts and loads without suffering permanent deformation or fracture. For example, it would be of great value to have glass with greater elastic properties. The "elastic metal" produced by the University of Michigan could revolutionize many areas of science, especially for electronics.

Metal alloys and reordering of crystalline lattices is a very promising field, since simple modifications in their compositions and structures can significantly modify their properties.

CERAMIC

The word ceramic comes from the Greek *keramos* and means "thing – clay – burned" was one of the first materials mastered by man, whose first pieces found date back to about 8,000 years B.C. As the name suggests, it is produced from the heating of clay under certain conditions, such as humidity, temperature and "baking" time.

The resulting material is formed by varying amounts of metals and non-metals linked by ionic and covalent bonds, its crystallographic structure generally has many faces and its atoms do not slide easily as metals do, so they are more fragile than metals (there are exceptions), however they resist high temperatures – some reach a melting temperature greater than 4,000 °C such as

69

Hafnium Carbide – higher than the resistance of many metals, which melt at lower temperatures.

Ceramics can be broadly divided according to scheme 4, being one of the most abundant groups with representatives ranging from stones to innovative synthetic superconductors capable of replacing metals.

Ceramics do not conduct heat and electricity well – with the exception of a small group that reach the condition of superconductors, when subcooled – and have high elastic modulus, a factor that contributes to the fact that they are mostly brittle and little conventionally used for the manufacture of springs and related artifacts.

Scheme 4. Basic classification of ceramics.

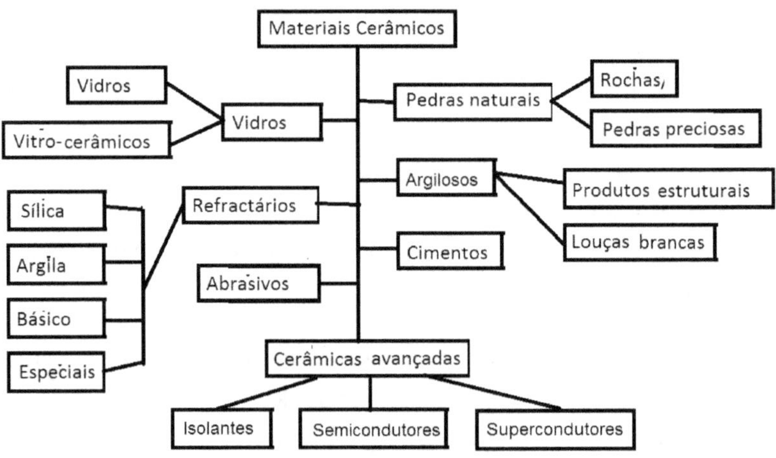

Source: author himself

The character of the bonds – whether predominantly ionic or covalent – has a great influence on the atomic arrangements and physicochemical properties of ceramics. However, the control of their elastic properties has been shown to be much more complex and difficult than for metals. There are no efficient methods for raising its elastic properties so far, although some such as alumina can replace some metals.

Even when referring to modern ceramics, such as dental and the electronics industry and special turbines in general, its elastic moduli still exist. They are also modified by temperature, some gaining significant magnetic properties.

POLYMERS

Polymers are organic or non-organic macromolecules, and can be of biological, non-artificial or synthetic origin, formed by smaller structures called monomers. They have very varied characteristics and can be divided into at least two groups in terms of their ability to be molded, thermoplastic when they can be heated and reshaped, and thermoset when they cannot be melted and return to their original conformation.

As for their elastic properties, they can be divided into plastics, when they do not allow great distension from their original shape, and elastomers, when their elasticity is very high, that is, with very small elastic modulus, in general < 10 Mpa.

Plastics, when subjected to a load, react by stretching more than most metals, due to their lower elastic modulus, however, when the breakage of their elastic limit occurs before the occurrence of fractures, a process known as "plastic phase" occurs, where the deformation of the material is permanent and the continuation of this can lead to breakage, The phase between breakage and plastic phase is known as the yield limit and, although it exists in other materials, in plastics it is very characteristic and notorious.

Polymers can be broadly classified according to diagram 5. It is seen that temperature is of crucial importance to them, especially for maintaining their structure and conformation, whether organic or inorganic, their elasticity depends on the mobility of their chains. Some are very dependent on moisture, especially biomaterials, to maintain their structure and elastic properties.

Scheme 5. Overall classification of polymers.

72

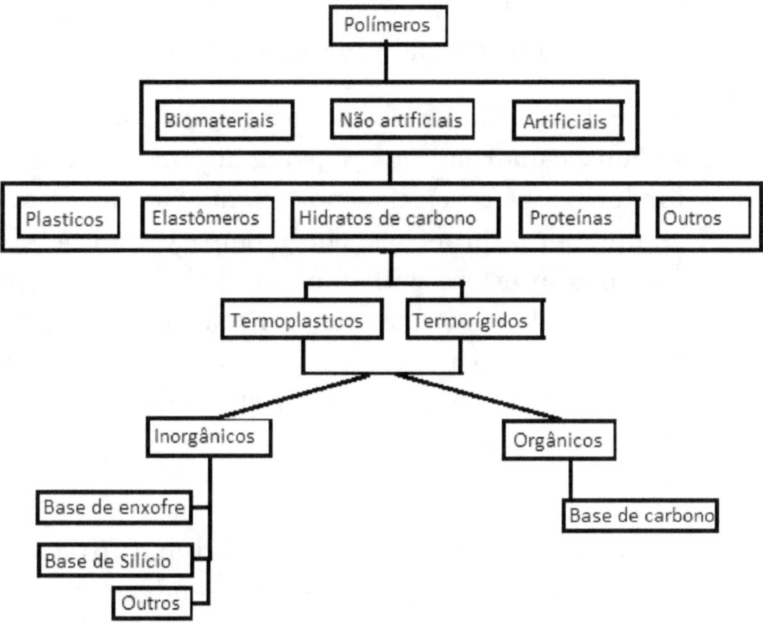

Source: author himself

Polymers such as some proteins and carbohydrates have a molecular structure that is extremely sensitive to humidity and temperature, and it is possible to manipulate the elastic modulus through their control. This is the case of chitin present in the exoskeleton of insects, a target of materials sciences that has been increasingly using it as an elastomer dopant, which if it loses moisture modifies its elastic modulus, which if it loses enough structural water it can be reduced to sucrose.

Although the temperature greatly influences the elastic modulus of polymers, the greatest control of it has been shown to be through the composition and density of the material through the manipulation of molecular chains and intermolecular forces.

Polymer molecules, due to their size, can assume various conformations, curling around each other and hindering the reversible movement of atoms. Factors such as number of branches, chain density, and hetero atoms can also interfere with the elastic behavior of polymers significantly, and since electrons are strongly attached to their bonds and atoms (more so than in metals) the chains can be more rigid, except when they are interconnected by weak bonds (electrostatic forces such as dipole-dipole, etc.).

In general, the more entangled and fixed the chains of a polymer are, the less elastic it tends to be and the less ductile as well. It can be brittle and brittle, also depending on the cohesive forces of resistance to impacts.

The fatigue of polymers varies from one material to another, however, they can be much more resistant than many metals, for example, high-density polyethylene, which can replace iron in some of its applications, however, it has little resistance to high temperatures.

Elastomers are polymers whose Yong modulus varies from 0.1 to 10 Mpa in general, they are very resilistes, that is, they return to their original shape quickly after suffering deformation and can deform a lot, some reaching more than 800% of their size. Its molecular shape is similar to that of plastics, however, it has intermolecular relationships that allow greater and easier displacement of molecular chains, the elastic limit is at the maximum point where the body can be stretched without the permanent rupture of these chains or their allocation in space.

When past the load demand, forces such as Van der Waals and intermolecular bridges "pull" the chain to its original conformation. This mechanism can be mathematically shown by the thermodynamic theory of rubber, which expresses through the second law of thermodynamics the reasons for it. Because, when stretching occurs, the amount of entropy of the elastomer is reduced and when the load is released, the entropy is reconstituted.

Thus, one can write the deformation L as a function of the partial derivation of you and S from the first and second laws of thermodynamics, writing (Eq.18). Being f the energy.

$$f = \left(\frac{dU}{dL}\right)_{T,V} - T.\left(\frac{dS}{dL}\right)_{T,V} \qquad \text{(Eq.18)}$$

In this way, a consequence to explain the flexibility of rubbers is that when they are stretched they acquire less entropy, as the second law of thermodynamics designates that the nature of things tends to chaos, the resistive energy pulls the molecules to the state of higher conditional entropy, with this, the original form of the elastomer, however, there is an entropy not conditional to the elastomer that tends to destroy it, the same one that tends to destroy all things.

When the energy is fully restored, the body is perfectly elastic and there is no substantial loss of energy in the thermodynamic process, which would be a contravention of Carnot's principle, therefore, elastomers tend to fatigue the greater the amount of deformations, because no transformation allows total conversion of energy into work, always to losses.

COMPOSITES

Composites are materials formed by a matrix, for example, polymeric, ceramic or metallic together with fibers or particles, these form a composite with properties superior to those of the two components separately and the elastic modulus is closely linked to the composition of the system and the matrix/particle percentages.

The control of elasticity can be carried out, in addition to the above, electrically or thermally, depending on the materials that form them, not the general rule for this due to the wide variety of mixtures of materials. However, it is possible to produce materials that are literally capable of "moving" through electrical discharges in the same way as muscles and, therefore, modifying their elastic potential in a localized and complete way.

Figure 6 shows the main classes of composites and used, where it can be seen that they are divided according to the type of aggregate, the shape and sizes of the aggregates and the direction and layers that are arranged in the matrix.

Some composite nanomaterials and even polymers can still be physically and structurally modified with temperature, so that their properties can be "mutant" according to ambient temperature, as well as their elastic modulus.

Composites are probably the class of materials that are most familiar to us, ranging from bones, muscles, wood, concrete to innovative synthetic materials capable of replacing muscles and other human tissues. However, the particle/matrix ratio is almost always responsible for the elastic potential and stiffness. The response to electrical impulses or thermal elevations depends on their composition, although they have a lower response capacity than other monolithic materials, such as steel, iron and aluminum.

Scheme 6. Main grades of composites.

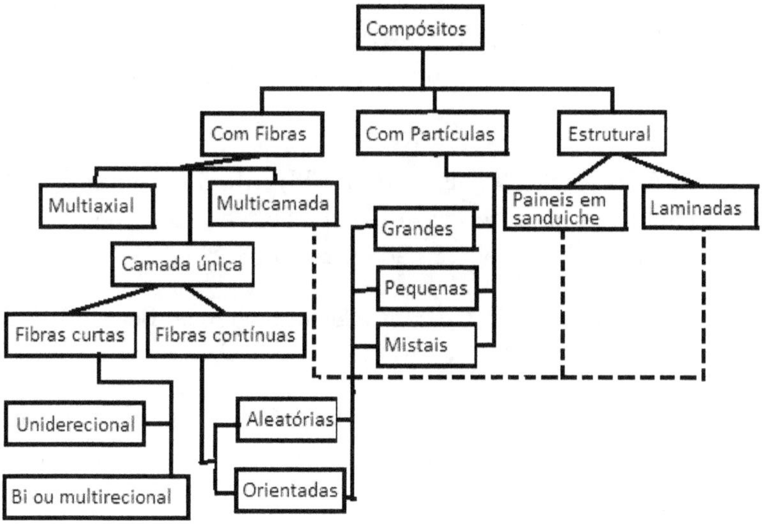

Source: author himself

It is important to emphasize that, as composites are very varied, their mechanical response to efforts is outside specific ranges, as occurs with metals, depending little on moisture for some types – generally synthetic derived from petroleum – and a lot for composite materials with a protein matrix and carbohydrates.

The microscopic structure of the composite is of fundamental importance, whether it is matrix or ceramic or polymeric particles, what actually determines its elasticity is the shape, concentration and arrangement of this aggregate. Whose best elasticity arrangement varies from one matrix (material) to another.

REFERENCES

BORES, A. P.; SCHMIDT, R. J.; Advanced mechanics of materials. Copyright Materials, United States of America, 2003.

CASSU, S. N.; FELISBERTI, M. I.; Dynamic-mechanical behavior and relaxations in polymers and polymeric blends. *Quim. Nova,* Vol. 28, No. 2, 255-263, 2005.

SHACKELFORD, J.F. Materials Science. Pearson Practice Hall, chap 1, 6th ed, 2008.

DE FIGUEIREDO, A. D.; Concrete with steel fibers. Polytechnic School of the University of São Paulo, handout, Department of Civil Construction Engineering, ISSN 0103-9830, São Paulo, 2010.

BOGAS, J. A.; Structure and behavior of materials: ceramic materials. Instituto Superior Técnico de Lisboa, Course Workbook, Lisbon, 2013.

Elastic devices: moles

All things are elastic, but not all things are springs. It is how elastic they are that defines them as such, and the character of the artifacts that we have as springs is relative to their application and purpose.

A spring is any solid object with the ability to deform reversibly with loads demands, within a certain range of elasticity. The springs can be used for work transfer, mechanical energy, impact minimization, surface adjustment, and wave production.

Figure 4. Examples of old springs and applications.

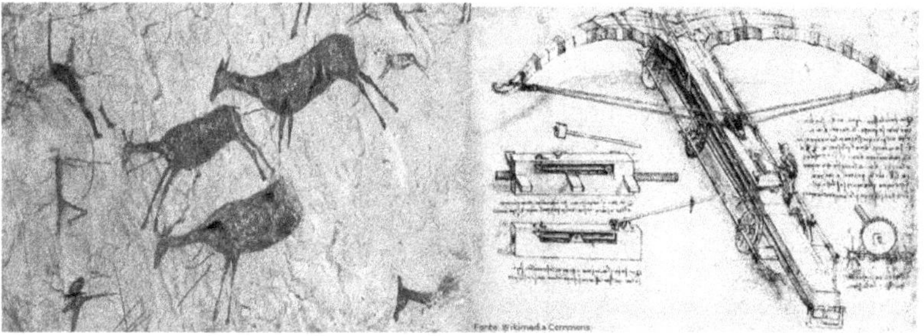

Source: FCETARCO (2016) and Da Vince.

The earliest molas have no specific historical dating, but cave paintings (fig.4) suggest that the earliest molas were the bows used to shoot arrows in the chipped stone age that lasted until 10,000 BC; This spring – bow – brought great social, warlike and dietary changes. Later, the arch was improved over time, being present in practically all cultures, especially in Egypt, Babylon, Persia, Syria, Greece and Rome.

Later, after the introduction of gunpowder in war weapons, the bow fell into some disuse, however, springs were always used, now inside the "bazookas" – old shotguns – and over time more and more among small objects such as watches and even carriages.

Springs aroused great interest again after the end of the dark ages, receiving theoretical foundation by Robert Hooke, who probably based on Galileo's work on mechanics developed what we now know as Hooke's law. Before he theorized them, it is likely that many springs were already being used during the Middle Ages and the ancient world, but there was no solid theory and mathematical description, at least there are no records, similar to what exists today for Young's module or some scientific specification.

Records about springs will probably be found soon, just as records about other technologies have been found, but it is likely not with the fidelity that Hooke did.

With the arrival of the modern age, springs began to be better studied for use in new machines, enabling a wide variety of spring-based products as well as new models, with applications ranging from musical instruments to vehicles and home properties.

Which, with the apogee of the industrial revolution, became of great importance for the manufacture of more resistant artifacts and conducive to the demands required. Now starting large-scale production and increasingly refined engineering based on specific metal alloys. Almost always their use was due to their resistance to loads, elasticity – damping and transmission of movement – and oscillative or vibrational capacities.

However, the best springs are those that combine the elastic properties of the forming materials with the structure and arrangement of the body that is constructed. Thus, an ideal spring is one that:

"To provide the best elasticity and resistance ratio for a given body disposition, satisfying the demands of loads requested for the use that was planned."

Spring is an object designed to satisfy the elastic demands of load requests, so the definition of all things are springs is not valid, it is more appropriate to say that all things are elastic, the other materials including natural ones with elastic properties can behave like springs. However, springs are synthetic or otherwise manufactured materials.

The most common springs are of three types: planers, helical and circular or plates. The less common are those in the form of levers and constant tension. Which have gained many applications being produced in plastics, metals, ceramics and other materials, presenting great utility in the automotive industry, household goods and consumption, electronics, aeronautics and warfare.

MOLAS HELICOIDAIS
This category of springs has no specific date of discovery or inventor, as there are varied reports about

their period of appearance. Although Robert Hook in the seventeenth century made one of the first reports of such an artifact and created his defining law for springs, these have probably existed since the Bronze Age as reported by Hooke himself1. Examples of spring shapes are shown in figure 5, which may undergo slight modifications in their body, but never leaving their helical shape.

Figure 5. Examples of helical mole body structures.

Source: Soberana molas (2016).

Springs have oscillating characteristics that can be described mathematically by the equation of motion. Being a simple harmonic movement, when the losses of energies in a real environment are not taken into account, the amplitude is constant.

$$y(t) = y_0 \cos\left(\frac{2\pi}{T}t + \alpha\right) = y_0 \cos(2\pi.f.t + \alpha) \qquad \text{(Eq.19)}$$

Where is the amplitude – maximum distension – constant of the phase of origin, T is the period. So we can derive equation 19 and find the velocity $Vy_0\alpha_c$ (Eq.19a), see that the velocity can also be given by equation 19b in angular form, which does not depend on the measurement point on the spring, informing the general velocity of the system or angular velocity.v_a

83

$$v_c = \left(\frac{2\pi}{T}\right) y_0.\,sen\left(\frac{2\pi}{T}.t + \alpha\right) = \left(\frac{2\pi}{T}\right) y_0.\,sen(2\pi f.t + \alpha) \quad \text{(Eq.19a)}$$

$$\omega = \sqrt{\frac{\left(\frac{2\pi}{T}\right)^2.(m_t)}{m}} = \sqrt{\frac{k}{m}} \quad \text{(Eq.19b)}$$

Equation 19 shows the shape of the acceleration for the oscillating system, so we can write that the acceleration is also proportional to the angular velocity, which to simplify the terms and reduce the number of variables, we can write it as the square of it with the product of the deformation y. *a*

$$a(t) = -\left(\frac{2\pi}{T}\right)^2 y = -\omega^2.y \quad \text{(Eq. 19c)}$$

Note that equation 19c demonstrates the second-order derivation of the executed function with respect to time. Therefore, one can find the force by applying Newton's equation, shown in equation 20, where m is the mass of the body suspended in the oscillating system and $my(t)_t$ is the specific mass, f is the frequency and T the period.

$$\sum F = \frac{dp}{dt} = \frac{d(mv)}{dt} = m.a = -\left(\frac{2\pi}{T}\right)^2 y.m_t = -\omega^2.y.m$$

$$\text{(Eq.20)}$$

However, remember that the mass of the spring must also be taken into account, after all it interferes with the oscillation system, so we can write the equation z in the form of equation 21, where we can make m_t as total or specific mass. Since $m_t = $, note that in this sum the influence of the mass of the spring is only a third on its behavior, but the mass of the body is integral. Where m is the mass of the spring and M is the mass of the body. This is very useful in dynamic calculations for the elastic behavior in vehicle suspension systems and so-called "rope" devices.$M + \frac{m}{3}$

$$\Sigma_f^{f.} f = -\left(\frac{2\pi}{T}\right)^2 \cdot \left(M + \frac{m}{3}\right) \cdot y \qquad \text{(Eq.21)}$$

We can also rewrite the equation p in the form we conventionally know by Hooke's law, replacing the terms mass and acceleration with the deformation suffered y and a constant of proportionality k, since the results for force can be known by the equation p, we can simply write the "squashed" by means of the constant k for any value of y in a given force f, arriving at equation 22.

$$f = -k.y \qquad \text{(Eq. 22)}$$

See that k corresponds to the specific mass of the system plus its acceleration. Thus, for any spring that performs simple harmonic motion under a given force, we can use equation 22 without the need for all the variables of equation 21. In this way we can write that the period of oscillation T can be given from the multiplication of 2π by the square root of the ratio of the specific mass to constant (Eq.23) and, as we know that the frequency is the inverse of the period and the angular velocity of the spring and proportional to the latter we can simplify as shown, where is the angular velocity.ω

$$T = \frac{1}{f} = 2\pi \sqrt{\frac{m_p}{k}} = \frac{2\pi}{\omega} \qquad \text{(Eq.23)}$$

Helical springs can also operate in sets, and equation 24 is the mathematical expression that demonstrates the relationships between the forces in parallel interaction, where occurs in conditions where the center of mass is explored, the sum of the elasticity constants k_n of both springs, giving rise to a specific elastic constant of the system k_e.

$$\sum_n^{n.} f_n = -y \sum_n^{n.} \left(\frac{2\pi}{T_n}\right)^2 \cdot \left(M + \frac{m_n}{3}\right) = -y \sum_n^{n.} k_n = -k_e y$$

(Eq. 24)

Thus, the kinetic energy of the set of parallel springs can be given similarly as it is calculated for conventional isolated springs, taking into account k_{and} since this fact replaces the elastic state in the equation with the specific constant. Accepting equation 25, to obtain the kinetic energy from the integration of the deformation y using the conventional equation of the kinetic energy E_c and using the elements of the equation of the acceleration of the oscillating system multiplied by the time t_p to convert it into velocity.

$$E_c = \frac{1}{2}\left(M + \frac{m}{3}\right) \cdot \left[-\left(\frac{2\pi}{T}\right)^2 \cdot t_p\right]^2 = \frac{3M+m}{6} \cdot v^2 \qquad \text{(Eq.25)}$$

$$E_p = \frac{1}{2}k \cdot y^2 \qquad \text{(Eq.26)}$$

Since the equation Eq.26 is applied to obtain the elastic potential energy, which when in an ideal system is free of energy losses, it can be assumed that the kinetic energy is equal to the work w that it can perform. The mechanical energy of a spring, on the other hand, can be

86

described as the sum of all other forms of energy in the body.

$$E_m = w + E_c + E_g \qquad \text{(Eq.27)}$$

Where w is the work or elastic potential energy, E_c is the kinetic energy of the spring, and E_g is the gravitational potential energy.

BODY CHARACTERISTICS: MANUFACTURING AND ANALYSIS

Although simple, coil springs have many characteristics that must be considered before being manufactured, so that the maximum use of the spring's elastic potential for its intended purpose is guaranteed.

We can equate the forces in a coil spring taking into account its bodily characteristics, so that any effort becomes unique to this category and if I can understand, how the relationship of other forces acting on the performance of a spring occurs, such as torque, shear, maximum safe stress limit, size and maximum diameter of a spring and characteristics of its forming wires. Thus, we can establish that the force can also be calculated by equation 28.

$$F = \tau.\pi.\frac{d_f^3}{K_w 8 d_m} = \frac{\tau.\pi.d_f^2}{K_w.8.l_m} \qquad \text{(Eq.28)}$$

Where is the shear stress, $d\tau_m$ is the mean diameter of the spring, π is the trigonometric constant, C is the curvature index, d_f is the diameter of the wire, Kw is the Wahl constant. By this equation it is possible to use characteristics of coil springs and, thus, predict any, by substitution of variables, characteristics unknown in an

87

analysis. Note that the torque T can also be calculated when taking into account the force F and the diameter D of the spring, considering the number 2 as a divisor as a function of the ends. According to equation 29.

$$T = F.\frac{D}{2} \qquad \text{(Eq.29)}$$

Therefore, to build a coil spring, fundamental aspects must be considered, such as: elastic potential and type of material, arrangement of the spring and turns, distance and maximum number of turns and size of the spring.

Tensile strength and material type

As seen in chapter 2, there are many materials that can be used to manufacture springs, but as for helical springs, it is necessary that the material has enough strength and elastic potential to be formed as a wire and be bent into spirals so that these spirals are as resistant as possible and can be produced with the greatest distance, resistance and the lowest possible thickness.

This characteristic is related to the intersection of several physical properties of materials, such as tensile strength, ductility – ability of atoms to slide over each other – impact resistance, elastic resistance, etc., where, the greater the elastic potential and the lower the plasticity of the material, the more suitable it is to build a spring.

Metals and their alloys are the most used materials for the manufacture of coil springs, followed by some polymers and blends, with ceramics being the materials with the least use in view of the improvement of techniques and the study of compositions presenting few advantages for the manufacture of springs, with rare exceptions.

In order for a spring to maintain its original shape and characteristics, the limit tension t (eq.30) must not be exceeded. The calculation to know the maximum load (eq.31) supported C_m, considers that the product of the maximum deflection, wire diameter $d\delta_{tf}$ and transverse modulus of elasticity G by the product ratio of 8 times the index of the spring and the number of active turns, generates the maximum value of force that a spring can support. Thus, the larger the diameter of the wire or the lower the index of the spring, the more resistant it tends to be, as well as the larger it is and G. F m being δ_t the maximum force on the spring

$$t_m = \frac{8.F_m.I_m.K_w}{\pi.d_f} \qquad \text{(Eq.30)}$$

$$C_m = \frac{\delta_t.d_f.G}{8.I_m^3.N_a} \qquad \text{(Eq.31)}$$

If the spring, when it is required by a certain effort, is allowed to reach solid length – when all the turns are against it – it can lose its properties, as it changes its elastic modulus. Therefore, a gap of 10% to 23% of the maximum deflection is left, so that there is no plastic deformation due to the breakdown of the yield limit. Equation 32 is for determining the deflection of the spring by the Castigliano method.

$$\delta = \frac{8.F.d_m^2.N_a}{d_f^4.G} = \frac{8.F.I_m^3.N_a}{d_f.G} \qquad \text{(Eq.32)}$$

The equation establishes, by means of data such as the number of active turns Na, applied force F, transverse modulus of elasticity G, wire diameter d_f and spring diameter dm or spring index I_{mm} a value that corresponds to the flattening that can be applied. The use of the same is of great importance in the calculation of safety for automobile springs.

Equally interesting for industry is the obtaining of the elastic constant k of the spring through the bodily characteristics of the springs. Equation 33 demonstrates how we can find the constant or the other inherent variables mathematically and thus determine whether a spring built with certain measurements and material will meet the engineering demands that are designed.

$$k = \frac{d_f . G}{8 . I_m^3 . N_a} \qquad \text{(Eq.33)}$$

Arrangement and thickness of spring and turns

The ideal maximum thickness of a coil spring is calculated based on the wire diameter of its turns, the distance between the turns, and the characteristic of its material. In this way, the springs are designed so that there is no flow – effect of plasticity – which we can calculate through practical studies and scientific literature according to the wire diameter resistance graph for each material, graph x, where T_L is the limit stress.

In this perspective, the index of the spring I_m (eq.34) is the quantity that relates the diameter of the spring to the diameter of the wire. It is this index that helps the calculations to obtain how large a spring can be while maintaining its resilient characteristics. This measurement is necessary because the curvature and inclination of the turns is directly related to the shear stress. As shown in equation &, where the relationship between all the bodily items of the spring, as well as with some of its mechanical characteristics.

$$I_m = \frac{d_m}{d_f} \qquad \text{(Eq. 34)}$$

In this way, there are limits that must be obeyed to ensure greater resistance and durability of the springs in relation to their maximum diameter, so the indexes are established according to the purpose for which they are desired, as for example, in the case of clutch springs which, in general, have a limit of $I_m = 5$. Because, if a spring has a diameter larger than the ideal envisioned by its index, the greater the chances of losses occurring, either due to plasticity, low elasticity or the absence of the desired response to a load.

The resistance of the wire and the interactions of the forces acting on it can be related by the Wahl factor W_h for variable (eq.35) and static (eq.36) stress and, as the curvature for the latter is not considered, the first term of the equation is equal to 1. This factor considers the index of the spring I_m, a correction for "shear effort" – the second side of the equation – and the natural stress – the first side of the equation – due to the curvature of the turns, which is greater in turns with less curvature and more rigid.

$$W_h = \frac{4\,I_m - 1}{4\,I_m - 4} + \frac{0{,}615}{I_m} \qquad \text{(Eq. 35)}$$

$$W_{h'} = 1 + \frac{0{,}615}{l_m} \qquad \text{(Eq. 36)}$$

The Wahl factor is responsible for determining, in a way, the maximum limit for resistance/characteristic control that a spring of a given structure and material supports. Some use this dimensionless quantity to "replace" the shear stress.

It is necessary to highlight that equations V implicitly take into account that the greatest stress occurs in the internal region of the spring, and therefore, the diameter of the curves of the turns are determinant as the resistance of the spring, so the Wahl factor "replaces" the shear, for this it takes into account a safety limit for stresses.

The physical size of a spring depends on experimental tests and is directly related to the index of the spring. However, it is considered that a spring in exercise can have the maximum and minimum elongation, so that it can have enough firmness not to buckle when required by a force, better elasticity/length ratio and good durability. Thus, when calculating the minimum size (Eq. 37) a percentage of safety is considered, usually ranging from 0.1 to 0.23 whose product of it with the maximum deflection is added to the length of the closed spring or solid length.$\delta_t l_f$

$$l_m = l_f + p_s . \delta_t \qquad \text{(Eq.37)}$$

$$l_{mx} = 4 . d_m \qquad \text{(Eq.38)}$$

The maximum size of the spring, we can calculate according to equation 38, where the main factor is the diameter of the spring and a numerical value whose product expresses the point at which the spring begins to buckle and present other deficiencies.

Distance and number of turns

The total effective number of turns is obtained by subtracting the first and last turns from the general quantity, however it can vary with the type of spring tip. The number of active turns can be given by means of equation 39, which lists all relevant entities so that the turns actively participate in the elastic process. Such as wire diameter d_f, spring diameter dm, force F, transverse modulus of elasticity G and elastic constant K.

$$N_a = \frac{d_f^4.G.\delta}{8.F.d_m^3} = \frac{d_f.G}{8.l_m.k} \qquad \text{(Eq. 39)}$$

The turns are separated by an angle of inclination and therefore, by a minimum distance p when it is not under the action of a force. Which directly influence the other characteristics of the spring, so we must apply equation 40 and consider that the angle must be less than 12°. Let the spring pitch p – distance between two turns – be calculated by equation 41.θ_i

$$\theta_i = arc\ tg.\frac{p}{\pi.d_m} \qquad \text{(Eq.40)}$$

$$p = d_f + \frac{\delta}{N_a} + e_p = d_f + \frac{\delta}{N_a} + p_s.\frac{\delta}{N_a} \qquad \text{(Eq.41)}$$

Coil springs can take many forms, including a conical shape, so although many of the equations can still be used, it is necessary to highlight that for each model

93

within this category it requires physical and mathematical singularities in the construction process.

FLAT MOLES

This category of springs is widely used and, as seen, they were probably the first springs used by man, and can be built in various materials, it is estimated that the first ones were made of wood – although it is believed that they were cylindrical and only later improved to a flat shape –

.

Figure 6. Examples of flat mola types.

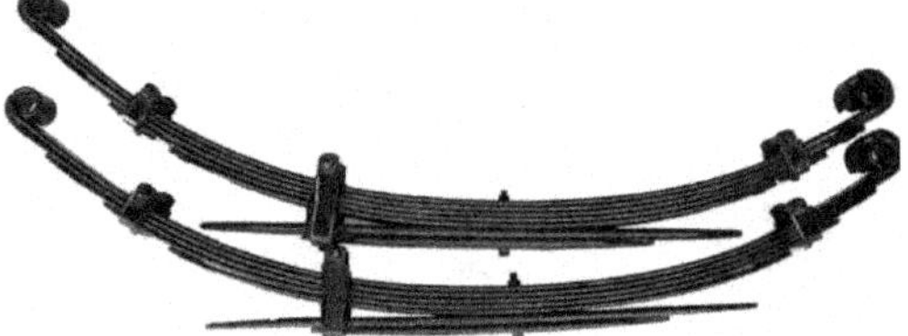

Source: Industrial Mechanics (2016).

Being widely used in the automotive industry, toys, electronic items and musical instruments. Figure y exemplifies this category.

The general equations for the behavior of springs can be easily employed for plane springs, but to obtain force and other quantities in a specific way, we can make mathematical arrangements.

These springs today have a rectangular section – for the most part –, length according to the determined index and variable width. The same rules are followed as the helical fibers regarding the distribution of force in their body, with the internal fibers being the ones that suffer the greatest tension and the external fibers suffer the greatest traction when a force is applied to them. Equation 42 shows that the stress does not depend on the series, arrangement, or arrangement of the spring, but depends on the rectangular cross-section s s_r, the radius of the thickness y, and the force F and the cross-section length of the spring L.

$$t_s = 0.\frac{y}{s_r} = F.L.\frac{\frac{h}{2}}{\frac{b.h^3}{12}} = 6.L.\frac{F}{bh^2} \qquad \text{(Eq.42)}$$

Where t_s is the tension, L is the length, F is the force, b is the width of the spring, and h is the thickness. Note that in this case a uniform spring is considered, if the spring has a variation in size/shape of section and thickness, we can use equation 43, so that we divide the body into sections so that we can calculate the elasticity variables in them individually.

$$t_s = 6.F_m.\frac{L_x}{w.t^2} \qquad \text{(Eq.43)}$$

Where w is the base and t is height. Note that the maximum tension takes into account the maximum force F_x that the spring can withstand.

The deflection or deflection of the spring – how much the spring bends – depends on the series that the springs are arranged, as we can see in equations 44 for calculations with elliptical series, from one to two quarters and 45 for obtaining deflection of elliptical series springs. Note that the deflection has changed significantly, this due to the superposition of the constants, notice that the numerical value has doubled for an elliptical series.

$$\delta = 6.\frac{L^3 F}{E.b.h^3} \qquad \text{(Eq.44)}$$

$$\delta = 12.\frac{L^3 F}{E.b.h^3} \qquad \text{(Eq.45)}$$

The maximum force F_m that a flat spring can suffer can be calculated using equation 46, in which the deformation – maximum deflection – and the constant k must be considered, which, as seen above, can be related to the body characteristics of the springs. Therefore, we can rewrite them to satisfy the needs of visualizing the best spring design of a given material by substituting k of equation 46 for equation 47 and also for equation 44 or 45, depending on the cross-section. $\delta_m \delta_m$

$$F_m = k.\delta_m \qquad \text{(Eq.46)}$$

$$k = E.b.\frac{h^3}{6L^3} = \frac{F}{\delta} \qquad \text{(Eq.47)}$$

The elastic constant can also be calculated based on equation 47, which is Hooke's law with spring body details for force and strain or deflection.

These springs are generally coupled to several others, according to the need of the desired purpose, the arrangement allows to increase the elastic potential of the springs – that is, of the set –. To do so, it is usually adopted that the system is a compact solid or a spring with "n" blades, so for the first case equation 48 can be considered. For maximum deflection and 49 for maximum tension.

$$\delta_u = \delta_t = 4.\frac{L^3 F}{E.b.h^3}.n^2 \qquad \text{(Eq.48)}$$

$$t_s = 6.F.\frac{L}{w.t^2}.n \qquad\qquad (Eq.49)$$

Note that, for the deflection considered, it occurs as a function of the square of the number of blades, while the strain occurs only as a function of n. Thus, the greater the number of blades, the greater the stress supported and the greater the deformation that the spring can suffer without breaking. However, being together represents a loss of energy due to friction, so that there is a damping of the force to be restored.

If we consider the spring as a solid body, we can summarize its characteristics by the calculation of only "one" blade. Note that in this case, we can use equations 43, 44 and 45, with small variations for the type of conformation and thickness, so that to calculate the deflection of a spring of variable thickness we can use equation 50 for parabolic series.

$$\delta = \frac{2F.L^3}{3E.}.\frac{12}{bh^3} = 8F.\frac{L^3}{E.b.h^3} \qquad\qquad (Eq.\ 50)$$

Note that being any spring produced with the intention of serving as a shock absorber for large loads, the best option may be the coupling, which represents profit, not only due to its simplicity, but also for its natural ability to lose part of the energy used to deform it in the friction between the blades.

Flat springs can have several applications in addition to relieving a force, such as the reeds of wind musical instruments, which are mostly flat springs made of wood or other material, with the objective of producing sound waves through vibration and rarefaction of air at given frequencies.

Therefore, given their potential, as well as we can calculate the vibration frequency, speed and acceleration of coil springs, equation 51a can be used to calculate the natural vibration frequency for a Euler beam and, from then on, have an idea of the vibratory behavior of a flat spring.

$$\omega = \frac{1}{(\Delta x)^2}\sqrt{y}.\sqrt{\frac{GI}{\rho A}} = \frac{1}{\left(\frac{L}{n}\right)^2}\sqrt{y}.\sqrt{\frac{GI}{\rho A}} \qquad \text{(Eq. 51a)}$$

Where is the ratio of the size of the spring L to the number of nodes n, then is the distance between the nodes. Since the natural frequency form, G is the modulus of elasticity, the moment of inertia that is different for each type of spring support, is the specific mass and A is the cross-section of the spring.$\Delta x y I \rho$

$$\lambda_i = \alpha^4.(\Delta x)^2 = \frac{\rho.A.w^2}{E.I}.(\Delta x)^4 \qquad \text{(Eq. 51b)}$$

Notice that equation 51a for natural frequency of vibration arises from equation 51b of the form of frequency of vibration. Remember that this method was developed by finite differentials, both equations can be simplified to meet the demands of boundary conditions for various types of support and quantities of nodes, but we will not go into detail about this, as it goes beyond the desired scope, for more information see the references.$\omega \lambda_i$

MOLAS BELLEVILE

This category of springs was patented by Julien François Belleville, Paris – France, 1868. Having been invented, according to the inventor, the previous year by himself, springs are small conical washers, one of their main advantages is their low deflexibility and high mechanical resistance. Its applications are varied, but in general they can be applied to valve systems, in vehicle engines and actuators, pins and bolts of high forces and prices.

Figure 7. Examples of billevile molasses.

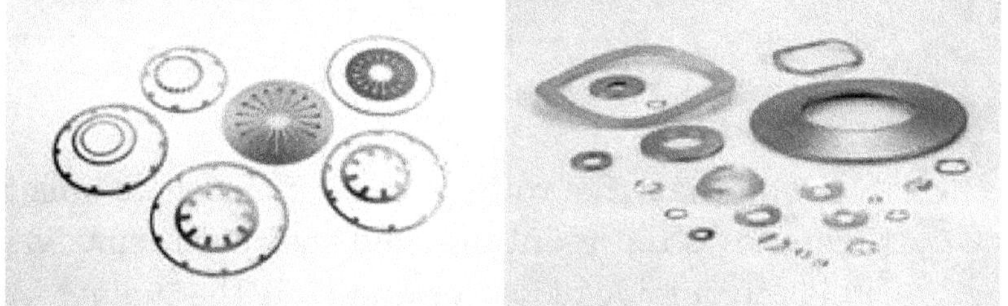

Fonte: Associated springs (2016).

These springs (fig. 7) have simple characteristics and their strength depends on the strength of the material. Note that for conical springs, the distribution of forces occurs in a way that is pertinent to their shape, and the low deflection has to do with this characteristic and the modulus of elasticity.

The calculation of the specific force of Belleville springs is given by equation 52, where the force F can be calculated considering the transverse modulus of elasticity G, the height of the spring h, the deflection, the outside diameter D, the Poisson coefficient v, the height of the spring h and the thickness of the spring t. Note that although simple, the calculation for its variables can be a little more complex than for flat and helical springs.δ

$$F = \frac{4\delta.G}{K_i D^2 (1-v^2)}\left[(h - \delta)\left(h - \frac{\delta}{2}\right)t + t^3\right] \qquad \text{(Eq. 52)}$$

Note that Ki in conical springs, represents the spring's elastic constant which can also be found by equation 53 in a specific way to Billeville springs. Notoriously, this constant can be obtained in a simpler way, however it would not be related to the desired characteristics of the spring.

$$K = \frac{6}{\pi.Ln\left[\frac{D_0}{D_i}\right]}\left[\frac{\left(\frac{D_0}{D_i}-1\right)^2}{\left(\frac{D_0}{D_i}\right)^2}\right] \qquad \text{(Eq.53)}$$

Note that the ratios between brackets inside the parentheses represent the solid body of the spring, where D0 is the outside diameter and Di is the diameter of the washer hole, so the result is a fraction to multiply the right-hand side of the equation, which is the trigonometric constant, L is the length and n is the number of turns.π

To calculate the maximum stress that Belleville springs can undergo, that is, the force required to make them flat can be given by equation 54. Whose variables already mentioned above, have a great influence on its maximum load capacity, notoriously the thickness of the t-spring is due to the diameter D, so no matter how large and thicker the spring is, you can find the elastic constant of a spring made of a given material – usually

Chapter 4

metals and their alloys – however, the equation can vary according to the specifics of the spring or its body.

$$F_m = \frac{4.E.h.t^2}{K_i D^2 (1-v^2)} \qquad (Eq.54)$$

The Belleville springs patented by their inventor have undergone many improvements and adaptations, some do not belong to this category, such is their mischaracterization. In this perspective, it is necessary to research for means of characterization for the springs that appear daily in the markets by experimental means and mathematical design in order to ensure safety, quality and technological development.

REFERENCES

ARCANJO, E. P; Characterization of spring fatigue behavior. Technical University of Lisbon. Master's dissertation in mechanical engineering, Lisbon – Portugal, 2008.

ALMEN, J. O.; LÁSZLÓ, A.; The Uniform-Section Disc Spring Trans. ASME 58, (1936) p. 305 to 314.

CORREIA, V.F.; Machine Organs: Sizing of coil springs. Infante d. Henrique, marine machinery engineering department. Oeiras - Portugal, 2005.

DAVET, G. P.; Using Belleville Springs To aintain Bolt Preload. Solon Mfg. Co. 2001.

DUBEY, H.K.; BHOPE, D.V.; TAHILYANI, S.; Singh, K; Effects of Slots on Deflection and Stresses in Belleville Spring. The International Journal of Engineering And Science, Volume 2, Issue 3, Pages 43- 48, 2013.

DOS SANTOS, A. A.J; Machine Elements 1: coil and flat springs. UNICAMP, course workbook, mechanical design department. Campinas, 2001.

MORO, N.; Machine Elements: Cool. Apostyle IFSC-department of metal mechanics, Florianópolis, 2015.

UNIVERSIDADE NOVA LISBOA. Determination of the elastic constant of springs. Handout, Department of Physics, Lisbon-Portugal, 2004.

__________; Sovereign springs. Manufacturer of springs, available in http://www.soberanamolas.com/ viewed on 06/2016.

__________; Associated springs. Manufacturer of springs, available in http://www.asbg.pt/produtos-servi%C3%A7os.aspx viewed on 06/2016.

__________; Industrial Mechanics. Manufacturer of springs, available in http://www.mecanicaindustrial.com.br/388-uso-das-molas-de-lamina/ viewed on 06/2016.

Da vince, L.; Sketch of Vince's Leonardo Beast. Image gallery of the Paraná Department of Education. Available on http://www.filosofia.seed.pr.gov.br/modules/galeria/detalhe.php?foto=44&evento=4 viewed on 06/2016.

FCETARCO ; From hunting to sport: the millennial history of the bow with arrow. Available at https://arcoeflechace.com/tag/historia-do-arco-e-flecha/viewed on 06/2016.

What are dialectical-thriving systems

Dialectical-thriving systems have always been with us, but we do not always recognize them as such. They were ignored and reduced to any other, even with the incredible and singular characteristic of having their elastic constant reversibly modified.

This category of system was initially created from an "in site" followed by observations and experiments with a money alloy (elastomer) and a straw (plastic), submitted to approximately 200 mechanical tests of longitudinal and transverse bending. Subsequently, the concepts were taken to the theoretical level of study of elastic bodies, hypothesis raising and rigorous experimental tests with engineering material.

Dialectical-thriving systems allow reversible variability of the elastic modulus of a body. Therefore, they disrespect the principle of invariability of the elastic modulus, in which at the same temperature the elastic constant is invariable within the limits of elasticity of a body.

From this perspective, two basic conclusions can be drawn from the previous paragraph, the first is that the principle was broken through this new theory and the second that there was *a relativization of the principle.* In this bias, it seems better to us to affirm that dialectical-thriving systems are exceptions to conventional ones —

simple systems – in which, in their case, the principle of invariability of the elastic modulus is relative.

The dialectical-thriving systems correspond to a new type of elastic system, and can be considered a new category of springs. It is often said that dialectical-thriving systems have always existed, however, the springs belonging to them were created.

The study began with the author's disagreement with an observed phenomenon that instigated the need to explain, measure and expand it. This phenomenon was first noticed in a small plastic cylinder with an elastic thread inside it and occurred whenever it was tensioned under different forces, noting that the elastic constant of the system seemed to vary.

In this perspective, experiments were carried out, through which, from the scientific literature and logical-mathematical knowledge, principles, foundations were built and means and objects were deduced to characterize and form systems with such capacity, which are called dialectical-thriving systems.

It is clear that the research is not intended to break the laws of classical physics, but to use them to provide data and information that are still hidden, thus providing new technologies, even if for some scientists this is considered "totally relevant". However, it is believed that truths are often relative and temporary, always requiring study. In this perspective, the theory aims to make a contribution to expand the human domain and knowledge about complex elastic bodies.

The term dialectic-thriving was coined in honor of the knowledge of a psychology professor, who stimulated the continuation of the research, *dialectic* is a mutual interaction that by causing alteration changes itself, with this, it is a system in search of balance self-organizing with each imbalance; *vigorous,* it means that which has power, great strength, vigorous. Therefore, they are systems capable of self-organizing to meet a load demand.

These systems, of a mechanical nature, modify their modulus of elasticity through the interaction of characteristic bodies (internal and external) where controllable changes in one of them cause modifications in the responsiveness of the entire system.

The term – modulable – indicates limited and reversible variation of the elastic constant, it was used so as not to cause disturbances or collisions of meanings in what tends to the understanding of the theme by third parties. Such was the dissatisfaction reported by some mathematicians when they read "theory of the variable elastic constant" who uneasily stated "if it is constant it cannot be variable".

The aim was to create and expand concepts, technologies and propose a guiding source for future research on a mechanism whose constant is modulable. Thus, the factors necessary for the production of dialectical-thriving systems and their main characteristics are outlined below.

FUNDAMENTALS OF MECHANICAL DIALECTICAL-THRIVING SYSTEMS

The parameters that govern mechanical dialectical-thriving systems represent an initial milestone in the study of this type of system, thus representing observations based on polymeric systems, which by theoretical extension are considered general, for the existence of these systems are necessary:

1. The existence of at least two elastic bodies classified as supporting body and supported body, the first being cylindrical;
2. The interaction of forces of semi-opposite senses between bodies, which are originated or maintained by tension/traction and elasticity;
3. The interacting forces must provide the formation of a single system, called the dialectical-thriving system;
4. The supported body must have the ability to change the intensity of the tension force on the support body, in which the constant of the system is modified thereby.

The proposed system can be analyzed through image 2, where it is observed that the interaction necessary for the formation of the system arises when the location of the internal body is considered ideal, that is, when its positioning favors the balance between the forces (b), originating an interaction Θ, in a simple way this balance seems to occur when the angle and position between the bodies allows the force exerted by the internal body to the support body opposes the consequences of tension that it causes to it.

Thus, if the same tension x is exerted on the cylinder in three different situations as shown in figure 8 in A, B and C it is observed that in B the equilibrium was formed and the interaction between the forces does not allow the system to deform perceptibly, then it is said that the system is dialectical-powerful, in A and C the system is not possible as below, because the angle and position favor a deformation in the system, not just a change in its properties.

Figure 8. Analysis of a body under the action of tension.

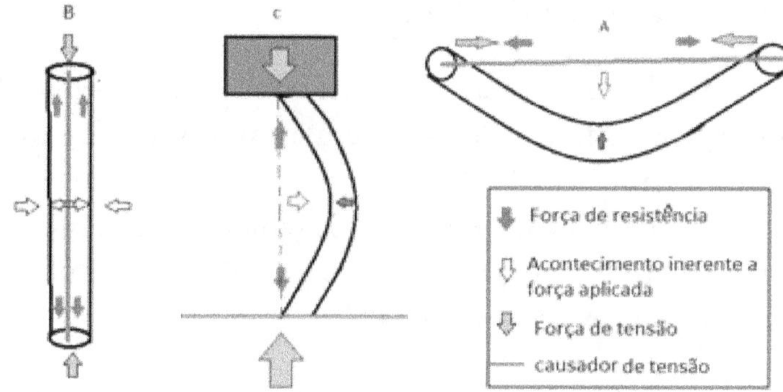

Source: author himself

The system analyzed in B is a polymeric dialectical-thriving system. Polymers are characterized as viscoelastic by the intermediate behavior between an elastic solid and a viscous liquid, which occurs as a function of the long molecular chains that constitute it and the intermolecular forces.

Therefore, the theoretical basis used as a starting point for the foundation of thriving dialectical systems is a mechanical structure made with polymers, with the

107

supported body being an elastomer and the supporting body a plastic. Thus, it is possible that bodies of other classes of materials have peculiarities regarding the formation of this type of system, however, they are in general terms as described in this chapter.

The reversible changes of elastic constant are given by a characteristic interaction between the forming bodies, this is distinguished from the related ones by the Greek letter Θ (theta). From this perspective, the interactions Θ, when properly manipulated, originate several elastic moduli in the same body or system.

Mechanical tests using Newton's laws reveal that the energy applied to the system remains in the system in equilibrium until another perturbs the system in ways other than the interactions Θ, which maintain it. Forcing the system to respond to elastic demand already influenced by the forces Θ.

In this perspective, the internal body, in order to be kept traction next to the support body, has to receive the same intensity of force that it applies in the opposite direction according to Newton's third law, however, the great differential of the dialectical-powerful systems is in the way the support body reacts to this "exchange of forces".

Even if the mechanical properties of the body that support it force it to react with a deformation to compensate for the action exerted by the internal body, the latter does not practice it, but varies its sensitivity to third loads, that is, its elastic modulus.

A fact that occurred as a result of the interaction Θ, which does not cause deformation of the support body, only energy storage and, consequently, change of some properties of the system, this change is the means by which the system complies with the aforementioned principles. Causing the system to enter an apparently static equilibrium from a macroscopic point of view, although at the molecular level a frenetic flow of energies is occurring.

The compression that the internal body is providing to the support body causes an approximation between the species that constitute it and the energy introduced forms a true "tug of war" between the compressive and repulsive forces of each molecule or chemical species, trying to assume new positions and/or resume the previous ones in both bodies that constitute the system due to the added force and energy.

It is important to remember that when it refers to traction it refers to the stretching of the internal body and when it refers to tension it refers to its action on the external body. That is why the term "tug of war" is used in allusion to this balance of forces.

This balance between both bodies provides them with the ability to resist and expand their capacities in what tends to loads, responding now as a cohesive and massive system, but with the ability to "manage" their elastic properties as the application of greater intensity of force is carried out via increased traction of the internal body, what is called modulation.

Experiments show that part of the force applied to the internal body is lost, either in the form of friction or by its natural fatigue, in addition to other characteristics of the system, such as the initial energy demand to form these systems.

EXPERIMENTS WITH DIALECTICAL-THRIVING SYSTEMS

Internal bodies that initially filled the entire internal surface of the supporting body, showed greater friction, which generates force and energy.

In springs, the position and angle at which the force is applied is very important for the valid exercise of its function. Therefore, in the systems under study, loads in two positions were considered: Horizontal and vertical. It should be noted that its properties in relation to dialectical-thriving systems are the same, although in a privileged way with the first test position.

The control of elastic potential energy in a dialectical-powerful system in the cylinder/elastic wire mechanism can clearly show the effects of the control of mechanical energy in transit, in the case exemplified in which a tense wire applies a force on the cylinder that holds it without causing deformation in it. The elastic potential energy, when added to the system, interferes with the value of its elastic modulus, making it more flexible.

In this way, in a simple and even didactic essay, it is possible to demonstrate the behavior of systems capable of modifying their elastic constant.

GRAPH 7. Comparison of data obtained in dialectical-strength systems under tension and their formers.

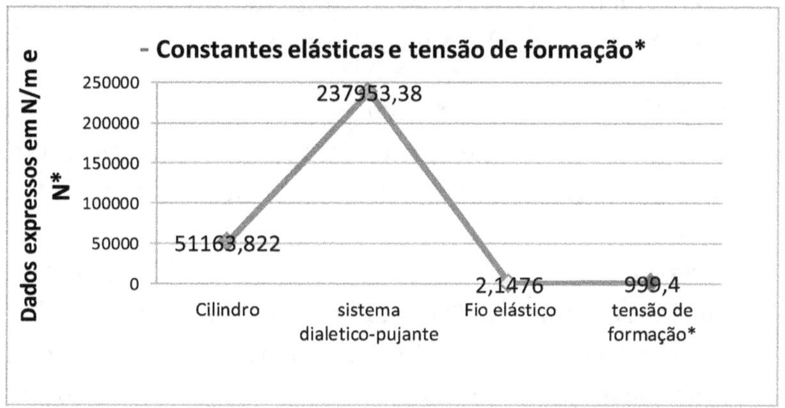

Source: author himself

Note graph 7, where the elastic constants of each body and the minimum formation stress of the system given in Newtons are shown, which is the traction that the external body must undergo for the system to respond cohesively as a single body to a load demand.

In the case exemplified, the formation stress is arbitrary, due to limitations of the method, however it is possible to construct graphically its effects on the elastic constant of the dialectical-strength system, these effects can also be observed in the increase of the elastic potential energy applied to the system, which forces a decrease in its elastic constant, as shown in graph 8.

The graphs reveal that, although it is only estimated, the formation stress exists and is located at the point where the bodies establish an interaction capable of forming a system, so for the capabilities of the method used, it was detected for the system in operation as 999.1 N, which may vary according to the properties of the bodies that compose it and the precision of the method.

The tension or energy of formation is the factor responsible for the insurrection of the system, a fact that can be noticed when there is a sudden increase in the elastic constant or modulus of elasticity in relation to the naked body or Hooke's system becoming a dialectical-powerful system. See that there is a great leap between ground zero and the first point, this interval determines the passage from the simple system to the dialectical-vigorous.

Graph 8. Analysis of the formation and behavior of the dialectical-thriving system under Epel and tension.

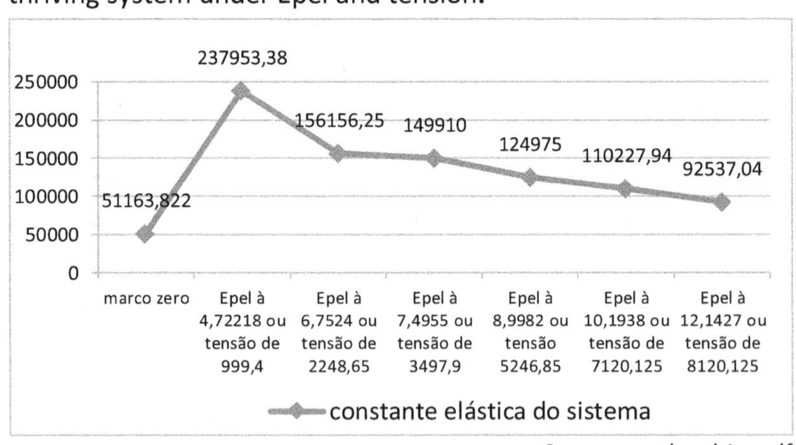

Source: author himself

112

Note that the tension of the internal body or supported body is given both in the form of force in N, and in elastic potential energy (Epel) in J and that the elastic constant falls the greater the force of the internal body.

Graph 8 reveals that the elastic constant of the system continues to increase with the decrease of the tension applied by the supported body until it reaches values below the *minimum formation energy*, being the point at which the supporting and supported body leave the characteristic interaction of dialectical-powerful systems and start to function as distinct bodies.

The zero point is opposite to the plastic phase, which is given as a consequence of the *flat point* or after the maximum voltage value. Thus, figure 9 illustrates how the various elastic constants are related, and that in the operating interval the dialectical-strength system works for each tension of the internal body with an elastic constant.

Figure 9. Schematic demonstration of characteristics of dialectical-thriving systems in operation.

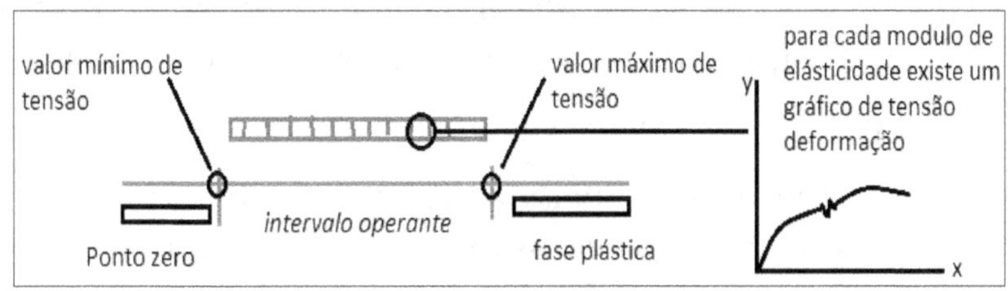

Source: author himself

Note that if the modulus of elasticity is modified, all mechanical properties are also modified, so in the same way that there is a minimum energy of formation, there

113

is also a maximum energy or stress of formation, which is given by the maximum value of tension that the internal body is capable of exerting on the external body without there being permanent modifications of its individual mechanical properties (of both bodies) that compromise the existence of the system.

The interval in which the system satisfies its demands without permanent modifications of its mechanical properties has been called *the operating interval.* This is different for each system and has agreement with the bodies that constitute them, they can be mathematically expressed on a scale of magnitude for classification test with values ranging from 0 to 100, it is expressed by equation 55.

$$i_{op} = \frac{k_{ma}}{k_{mi}} \qquad \text{(Eq.55)}$$

The closer to 100 the value, the greater the operant interval, when equal, close (> 80) or greater than one hundred the system is said to be *hyper-operating,* if less than 1 it is said to be *hypo-operating,* when it is in the interval between these two classifications it is said to be meso-operating.

In this formula i_{op} is the operating interval, K_{ma} is the maximum elastic constant supported by the system which is observed at the point where the minimum formation stress is evidenced and K_{mi} is the minimum elastic formation constant which is given at the point where the formation stress is maximum. In this way, it can measure the efficiency of the system by adopting "arbitrary" measurement scales, in the case of the system shown in graph 8, the I_{op} is 2.57 14, that is, this is a *meso operating* system, thus defining the size of a given operation interval.

Another important concept is *operability quantities*, i.e., the greatest number of times a system can be modulated or change its elastic constant. This quantity is divided into two categories, the quantity of temporal operability (GOT) and the quantity of existential operability (GOE), these are respectively the number of times in relation to the time that a system can be modulated and the total amount of times that the system supports modulations. Both are linked to the properties of the materials that make up the system and the characteristics of the system itself.

The GOT and GOE admit "arbitrary" and proper units that facilitate the understanding of the system, the first being expressed in a/m (number of changes/minutes), observed in equation 56.

$$GOT = \frac{a}{t} \qquad \text{(Eq. 56)}$$

Where *a* is the amount of modulations suffered by the system, obtained in values that are conventional to the end given to the system in a scalar or repetitive way, in a given time interval *t*, then the higher the GOT, the greater the capacity of the system to undergo modulations per minute in a range of changes.

The GOE is the maximum amount of changes that the system can undergo and is estimated only by the equation vc, where d is the "standard" strain used and n is the maximum number of times the system has undergone changes.

$$GOE = \frac{n}{d} \qquad \text{(Eq. 57)}$$

The systems in which the elastic constant is variable, inversely to the applied stress, were called hyprus, and the systems in which the elastic constant increases together with the tension are called haprocus, which have not yet been proven to exist, being only deducted. However, it is believed that both are governed according to the parameters illustrated in figure 9.

If the hypothetical similarities between these types of systems are proven, it is likely that only unit inversions are necessary to satisfy the haphazard systems in relation to the hybrid ones, since they are the opposite of the latter.

The nature of interactions in haptic systems is not yet known, but it is estimated that they are similar to interactions in water systems. Thus, there may be two options that are still little studied about dialectical-thriving systems, the aforementioned systems, and the advantages conferred by these two peculiar characteristics and their ability to resist through loads in elastic behaviors may be the great differential in the future use of variable constant systems.

Elastic moduli or elastic constants are basic and fundamental parameters for the application and description of materials. Several other mechanical properties are related to them, such as: rupture stress, yield stress, crack propagation under the action of thermal shock, toughness and others. Therefore , the increase in the control or manipulation of the modulus of elasticity can confer changes in its mechanical properties of a universal character in a given material.

REFERENCES

ARCANJO, E.P.; Characterization of spring fatigue behavior. Technical University of Lisbon, Master's thesis in mechanical engineering, Lisbon, 2008.

CHAGAS, G.M.P.; Machine elements. Discipline Workbook, IFSC, 3rd Ed., Jaraguá do Sul, 2009.

HAMROCK B. J., SCHMID S. R., JACOBSON B.; Fundamentals of Machine Elements; McGraw Hill; 2.ª edição; 2005.

MARTINS, W.C.; LAGO, W.W.S.; LOPES, J.S.; CUNHA, J.A.; SILVA, M.F.; Qualitative analysis of the elastic constant in Hooke's systems and in dialectical-thriving systems: constructing knowledge. Annals, VIII CONNEPI, Salvador, 2013.

117

Principles and technologies of synthetic dialectical-thriving systems

Dialectical-thriving systems have their technology still under development, but they can be applied by bicycle to earthquake-resistant bridges. Modulation technology can be mechanical or electronic, and can have integrated systems or just gears.

Dialectical-thriving systems are multifunctional and the technologies for them are still quite new in terms of their intrinsic components and their acceptance. However, the use of this technology in vehicles results in the possibility of providing it with maximum load capacity, without suffering deformation and with the same damping capacity.

The mathematical description of dialectical-thriving systems is defined based on the deformation of the internal body and its response in the elastic behavior of the system. Artificial systems of the types and forms set forth in this book and others that may appear as a product of manufacture, since in nature there are systems which regulate their elastic properties, but they are not springs.

Systems with variable elastic modulus can act in very different ways and capacities, however, there is a relationship factor that we can describe as a characteristic of the system, initially thought of as measurement error, it is now known that it is a value that describes the capacity of the system to self-regulate, being a dialectical-powerful constant called here the Cruz-Matos modulus, that is, the constant generated by the interaction of the forces Θ of attraction and repulsion of the internal (supported) and external (support) bodies after the minimum formation stress.

The Cruz-Matos modulus or dialectical-powerful constant, can be given as a percentage and represents both the deficiencies of the system and its active part, which in fact contributes to its existence.

MATHEMATICS OF DIETIC-VIGOROUS SYSTEMS.

Methods of measuring the elastic properties of springs essentially emerged with Hooke's law in approximately 1660, however, more accurate and efficient characterization methods are relatively modern and constantly changing. Around 1935 dynamic measurement methods emerged, other characterization methods have been studied since then.

In this perspective, the way to calculate the properties of dialectical-thriving systems to improve and better control them, in the sense of inserting them in engineering research and various studies. It has been demonstrating through its principle in a simple way that they work. The equations presented in this topic were developed by processing by the method of trial and error, until exhaustive tests that led to an equation capable of representing them.

The deformations of the dialectical-powerful systems can be calculated by means of the equation ddf, where is the new strain of the system by modulation, tx_{2ci} is the stress of the inner body, $t_{p\%}$ is the percentage of the stress that was lost k_{ci} is the elastic constant of the inner body, d_s is the

120

previous strain of the system, d_{ci} is anterior deformation of the inner body and x_1 is the anterior deformation of the system.

$$x_2 = \left[\frac{(t_{ci}-t_{p\%})^2}{k_c^2} - \frac{d_s}{d_{ci}}\right] + x_1 \qquad \text{(Eq.58)}$$

Note that, avoiding falling into a multivariable problem and without outlining the shape of the distortion curve, which is asymmetric, it is possible to determine the amount of deformation that the system will undergo under a stress of the internal body, based on parameters of the last deformation.

In this way, using the values of the constants in the previous chapter, the results of table 1 can be obtained, and if Hooke's law is applied, the energy conditions of graph 4 can be verified. Where clearly all the data show the similarities between the true values and the theoretical values.

The values obtained experimentally can also serve to evidence the increase in elastic potential energy and find the value of the elastic constant by the method expressed by Newton in the calculation of elastic potential energy, given by equation 59.

$$Epe = \frac{k.x^2}{2} \qquad \text{(Eq.59)}$$

Where K is the constant of proportionality, x is the deformation suffered, and Epe is the elastic potential energy of the system.

In this way, the values of the potential energy expressed in graph 8 are obtained, in relation to the

121

elastic constants, however, as there is imprecision of the measurement and other variables that were not controlled in the experiment such as temperature, fatigue of the bodies, etc., it is common that the true and calculated values vary in a similar order to that expressed in table 4, when calculated.

Absolute error/relative error %	Deviation from the mean Minimum/Maximum	Mean strain (cm) of the dialectical-strength system/position B at 1499.1 N	Deformation (cm) calculated by Eq.58
--	0,03/0,07	0,63	--
-0,028/-2,9166	0,06/0,04	0,96	0,932
-0,005/ -0,53	0,1/0,1	1,00	0,995
-0,053/-4,4166	0/ 0,02	1,20	1,253
0,0022/0,1617	0,05/0,04	1,36	1,3622
-0,013/-0,8024	0,04/0,08	1,62	1,607

Table 4. Analysis of the proximity between true values and calculated values.

Source: author himself

The causes of variation are pointed out by the poor precision of the measurements, the mechanical characteristics of the materials and the method used, which is alternative, as well as the materials. In this way, the pre-formula is put at the initial service of the dialectical-powerful systems, and there may be other means of calculating the degree or quantity of change suffered by the elastic modulation or the elastic constant of the system by adding energy to it.

When these results are shown at the energy level in graph 9, it is evident that the values corroborate, but there is still a small discrepancy, which needs to be studied for the application of corrective measures.

GRAPH 9. Analysis of elastic potential energy in dialectical-powerful systems media test with pre-formula.

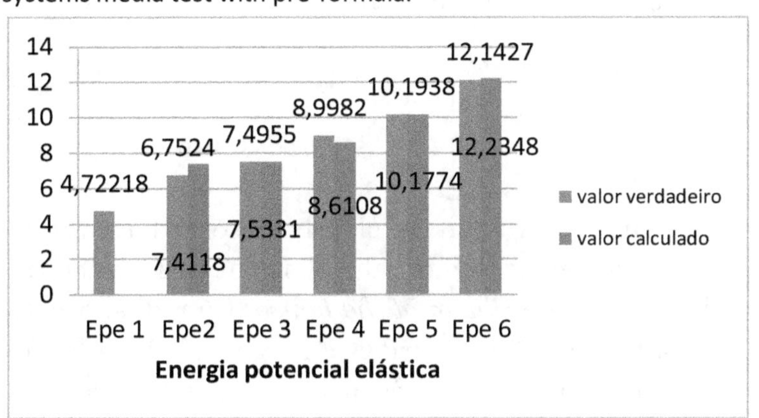

Source: author himself

The values obtained by the pre-formula in relation to the true value make it possible to partially control the system in view of the agreement between the values.

When there is a discrepancy in the calculation of the value of the new deformation, artifices are used to correct it. These artifices arise to the detriment of what has been called *the problem of the houses, a* phenomenon that occurs by the appearance or absence of decimal places – zeros – in front of the value found having this agreement with the true one, but it is much smaller or much greater, due to the *problem of the houses.* Generally, these decimal places are eliminated, which are extrapolative in accordance with the value to be subtracted from it, still within the parentheses.

See below an example of calculation with extrapolative values, in which it is possible to notice the relationship between the values. This problem may have arisen as a consequence of a principle not yet

123

demonstrated or some property of one of the system's forming materials. See that:

$$x_2 = \left[\frac{(t_{ci}-t_{p\%})^2}{k_c^2} - \frac{d_s}{d_{ci}}\right] + x_1 \rightarrow \rightarrow Xx_2 = \left[\frac{(1249,25-25\%)^2}{51163,822^2} - \frac{0,0063}{0,08}\right] +$$

$0,0063_2 = [0.000381 - 0.07875] + 0.0063 \rightarrow -0.072$

In this way, it is clearly observed that the value obtained, according to table 2, is not close to the true one, because the *problem of the houses* (in red) shows that the first value differs in order of magnitude by a lot from the second and especially from x1, to bring them closer one is sent an "order of magnitude" to the second value and given to it as many as necessary so that it is smaller than the first. In the case exposed, one more order in addition to the one donated, then one obtains:

X2 = [0.00381 − 0.0007875] + 0.0063 → 0.0093225.

Another example can be observed below, where another peculiarity of the *problem of houses can be seen, being:*

$$\rightarrow \rightarrow -3.5652 + 0.01 \rightarrow x_2 = \left[\frac{(5346,125-25\%)^2}{51163,822^2} - \frac{0,01}{0,0028}\right] + 0,01x_2 =$$

$[0,006105 - 3,5714] + 0,01\text{-}3.5642$

In this case, decimal places must be removed, multiplying the value by 10-n until it agrees with the value of x1 and the first value difference.

$$\rightarrow, x_2 = [0,006105 - 0,0035714] + 0,010,012534$$

If there is agreement between the first value and x1, as is the rule of the formula that the second value is less than the first, there is only the addition of three orders of magnitude (zeros) in the second value (10^{-3}) making it smaller than the first. In yellow highlighting, the difference between the value without the artifice and with it is observed. Previously shown, so that we can classify the first as not real.

In fact, the pre-formula is only a means of having control over the system, although it is partial it can offer a path for its improvement and temporary use, the variations occur on a small to medium scale, and can cause small changes in the values of the elastic constant of the system.

Based on graph 10, it can be inferred that the systems under study can be provisionally calculated by the formula already mentioned, however this has inconsistencies and disadvantages such as the fact that it does not allow the elastic constant to be calculated in the first modulation, because the values taken into account need another pre-established, that is, from the practical measurement of the first change it is possible to infer through the pre-formula the future values in the system.

The elastic constant can be modified (graph 10) and properly calculated when the values of x2 are considered, as seen in equation 58, and the imprecision is very small. This is promising, considering that it is not enough to know that the elastic constant varies, but it must be possible to control it so that it meets the required demands.

Graph 10. Test results and values calculated by the pre-formula under study.

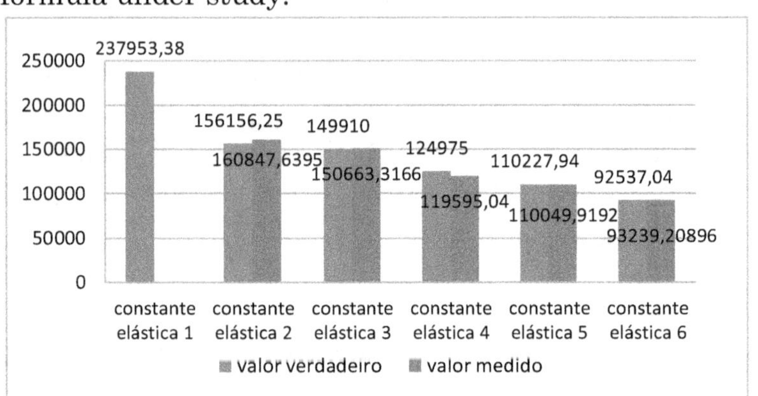

Source: author himself

Materials with more advantageous characteristics may offer better support for adjusting or recreating a more precise and cohesive formula. Graph 10 shows that the predictable variations of the modulus of elasticity present small inconsistencies due to the factors already mentioned, so that under given conditions the requirements of elastic moduli for the load in relation to the calculated value, do not allow the system to be used for purposes that require great precision without the insertion of a safety limit.

The *safety intervals* are used to characterize a percentage interval of precision of a given elastic modulus calculated by changing the elastic constant in relation to the true value, it shows a range where the probability of modulation safety is maximum.

For this, it is necessary to obtain previously from experimental data the maximum and minimum relative percentage error, then, a range of probable errors is

126

estimated, this probability allows the safety interval to be taken into account, which varies from one system to another.

MODELS AND ACHIEVEMENTS OF DIALECTICAL-VIGOROUS SYSTEMS.

It is evident that maximizing the interaction between the support body and the supported body, combined with the increase in the ease of modulation, can increase the efficiency of the system. Figure 3 demonstrates models to increase the efficiency of the system, which are improvements from the demonstrative model seen on the right of figure 2b, an example of a simple dialectical-thriving system.

In this bias, the dialectical-powerful systems can react in different ways to the stresses according to the force and disposition of the internal body on the supporting body, so systems with different structural arrangements and distribution of forces can be governed by a similar or different formula from the one mentioned above, and can also change the elastic constant in an inverse way to that observed in simple dialectical-powerful systems, being haphazard systems as already seen.

There is a great need for the various classes of dialectical-powerful systems to be tested in order to better understand them, it is estimated that there are at least four, as shown in Figure 10. These are respectively from left to right, called interak cruze, interakcircol, poliinterakhorize, monointerakhorize class.

Figure 10. Illustrations of varied models of dialectical-thriving bodies to obtain different interactions.

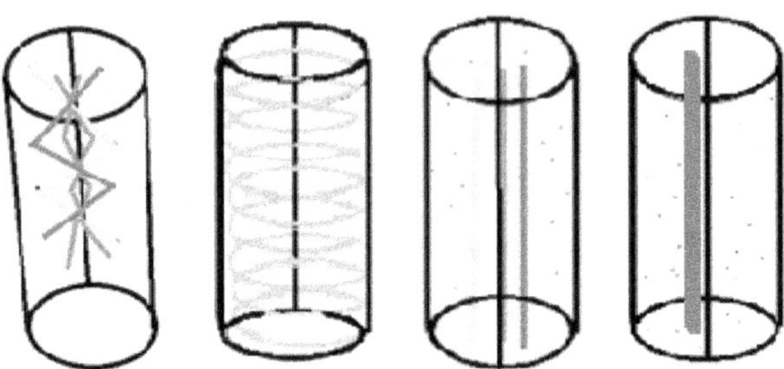

Source: author himself

All the dialectical-thriving systems that originate from the interaction between the external and internal bodies of cross-disposition belong to the first class of systems, the second class is given by all systems in which the disposition of the bodies or supported body is circular, semicircular or serpentine

The third class is very similar to the fourth, being different by the number of internal bodies, which requires greater care in what tends to the individual properties of these bodies and the tension applied to each of them to maintain the balance of the system, since the mechanical properties of internal bodies can vary, even having an ape nature, because they are elastomers.

The fourth class is the simplest and has the presence of only one internal body, which is responsible, through its tensions, for modifying the mechanical properties of the entire system. It was the base system of studies, from which the others were deduced.

128

The dialectical-powerful systems are also observed from the perspective of quantities and location of *points of eminent interaction* (which are said to be the place where the internal body exerts the greatest force on the external body). Thus, when the number *of points of eminent interaction* is equal to 2, the system is symmetric, when the system has more than two points and these are not in equilibrium, the system is asymmetric.

As for the location of the points of eminent interaction, when they respect an exclusive spatial geometric regime at the ends of the bodies, it is called the pical system, when the points are not exclusive at the ends, but are spread throughout the supporting body, the system is said to be apical.

The internal and external bodies are also given a classification, these are given intuitively according to their form and composition. Which, according to their shape, are said to be cylindrical, semi-cylindrical (all shapes that "resemble" a cylinder) and non-cylindrical, as for their composition they are said to be polymeric (when they are polymers) and metallic (alloys and related metals). These classifications are important to verify in more specific studies possible combinations and their efficiency for the desired purposes.

REFERENCES

MARTINS, W.C.; LAGO, W.W.S.; LOPES, J.S.; CUNHA, J.A.; SILVA, M.F.; The use of art and engineering to manufacture dialectical-thriving systems. Project and report, NPPGI, IFMA – Campus Zé Doca, partial report, Zé Doca, 2014.

BRITO, H.; Basic strength stroke of materials: pure tensile or compression. shear. Workbook of the Department of Engineering and Geotechnical Structures, University of São Paulo, 1st Issue, São Paulo, 2010.

L.H. Sperling, Introduction to Physical Polymer Science, John Wiley & Sons, New York, 2006.

Thomazi, Daniel A. F.; *Experimental and Numerical Analysis of Elastomeric Compounds*; Dissertation for obtaining the Master's Degree in Mechanical Engineering; Porto Alegre, 2009.

Introduction to the engineering of dialectical-thriving systems: the state of the art

All laws and elastic principles need to be applied to a material good to make it suitable to meet the multiple demands of human needs, which occurs through the construction of models of objects by the use of ingenuity.

In this chapter you will have basic notions of how to design pilot test models of a dialectical-thriving system, whose manufacture of prototypes will require technical and scientific knowledge, for this reason it is advisable to use an experienced person if you have little manual skill with ceramics and polymeric resins.

Simple dialectical-strength systems, as shown in the previous chapter, can be easily produced, since only a modulation mechanism is needed, which is a support to hold the elastomer inside the external body, so that it is possible to increase and decrease its tension.

If you want to produce a system in a more orderly way, you can manufacture the parts of the dialectical-thriving system using clay, polymeric resin with fiberglass and elastomers. This chapter exists to stimulate the production of works with the developed technology, so that it can be improved and expanded, ensuring the dissemination and stimulating the production of new prototypes.

PRODUCTION OF INTERNAL MOLD AND MODULATOR MOLD

Due to the uniqueness of the mold and modulator, these must be produced cautiously so that they meet the required needs. The molds are manufactured from clay.

The production of the mold from clay will require a lot of practice and attention from the researchers. The clay should be dried at a temperature of 30°C to 40°C, allowing only its partial curing, avoiding cracks, exfoliations and ensuring its ability to dissolve in water;

The clay will be sculpted according to measurements directly proportional to the inside of the stick. Which has 1.5 cm fittings for extenders, so cavities will be made in the mold in the format pertinent to these, the distance from one cavity to another, the diameter and size will be as per image 11

Figure 11. Molds for dietic-powerful systems.

Molde interno A

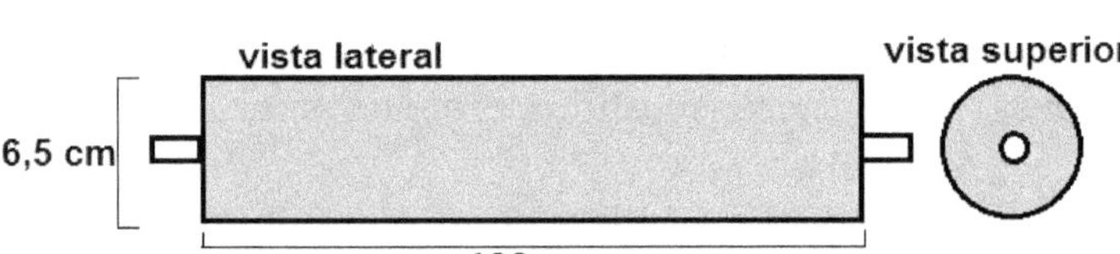

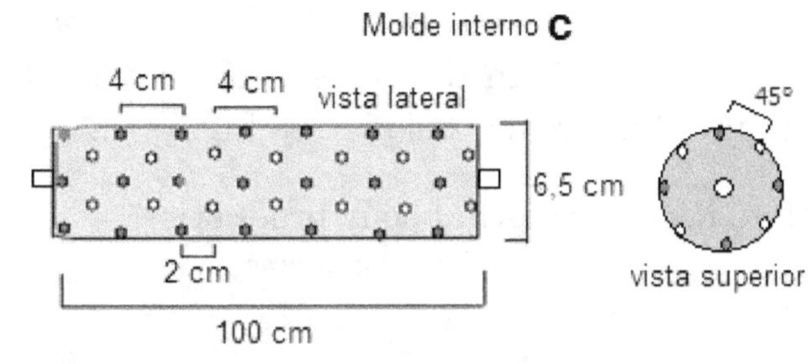

Molde interno B

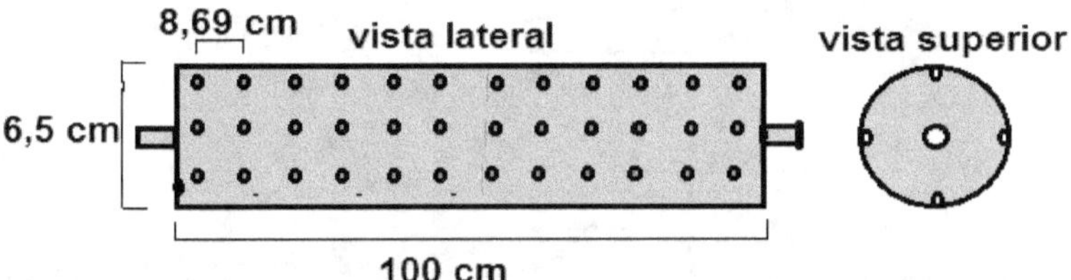

Source: author himself

ANALYSIS OF THE PHYSICAL BEHAVIOR OF EXTENDERS

This step is necessary in order to measure the elastic constant of the extenders and their degree of degradation, their resistance and useful life under the conditions to be submitted during the process and in the use of the product. The extenders are formed by several "wires" inside, which will be used as extenders of the rod.

The measurement of the elastic "constant" can be obtained according to the Hook formula, by applying a known force to the extenders or directly in a flexion test in a hydraulic press.

MANUFACTURING OF THE MODULATOR AND EXTERNAL BODY

133

This will be produced from fiberglass and polyester resin, will have fittings for the extenders and for the upper part of the rod, which will have mechanisms that allow it to move in rotations of up to 720°. The steps for its manufacture dry below. See Figure 12:

FIGURE 12. External body production.

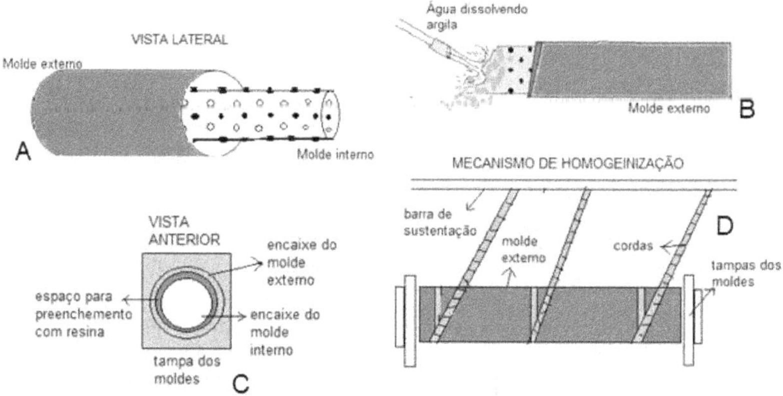

Source: author himself

1. Moisten the contact surface of the mold with release agent before applying the polymerizing solution;

2. Preparation of the polymerizing solution as obtained in item 4.4 of this project;

3. Use a suitable surface to add the curing machine to the molds;

4. After the polymerization of the substance, it will be removed from the molds of the pieces and possible imperfections will be repaired with sandpaper, brush, spatula and small amounts of resin;

5. They are assembled and made available for use;

6. Then the modulator is assembled and the functionality test is performed; See Figure 13.

Figure 13. System modulator.

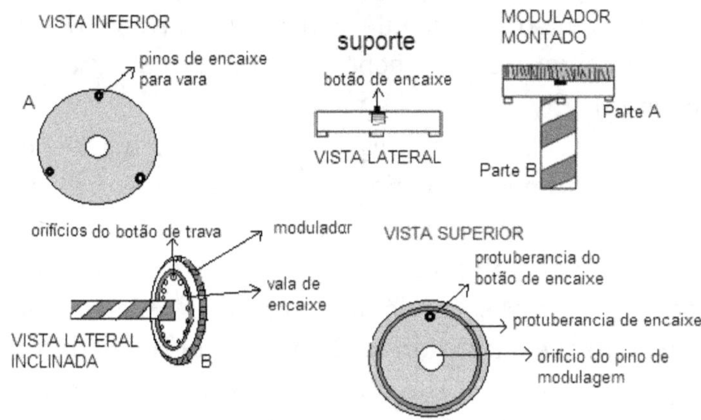

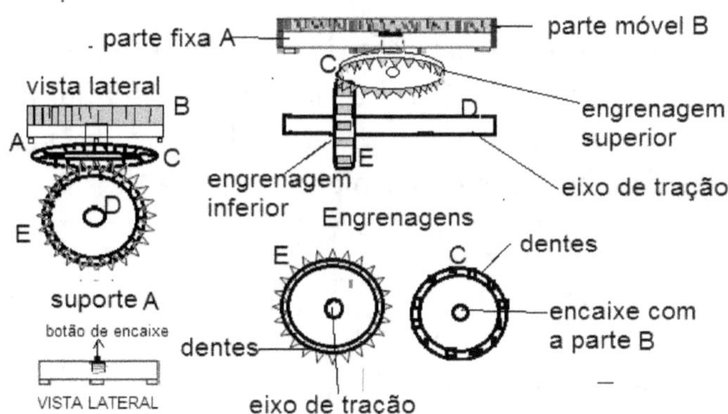

Source: author himself

ROD ASSEMBLY

This stage will require a lot of tests, dexterity and concentration, as a mistake can cause the loss of all the work, so two techniques were selected to be tested:

1. Cable alignment: this consists of introducing nylon wires into the cylinder and then pulling the elastic wire into the system;
2. Or like a spiral "needle" with nylon attached to it, then the extenders are attached to the thread and replaced mechanically;
3. Alignment to the magnetic force: with the help of a small metal needle, a magnet and the force of gravity, the extenders will be placed in their receivers;
4. Permanent alignment: during the manufacture of the rod, metal wires previously soaked in release agent are placed in the clay mold in the region of the extenders and after the drying of the rod, the extenders are attached to these that will replace them by means of mechanical traction.

After the proper positioning of the extenders, they will be lubricated with lubricating oil 30 and, after that, they will be attached to the modulator in the appropriate supports, being fixed with glue; Figure 14.

Figure 14. Scheme and structure of dialectical-thriving systems.

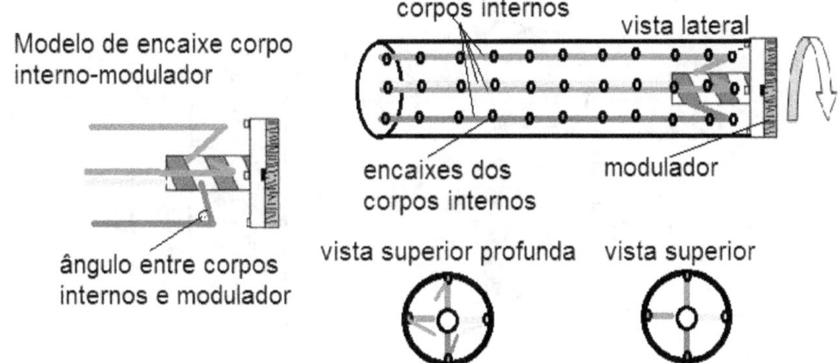

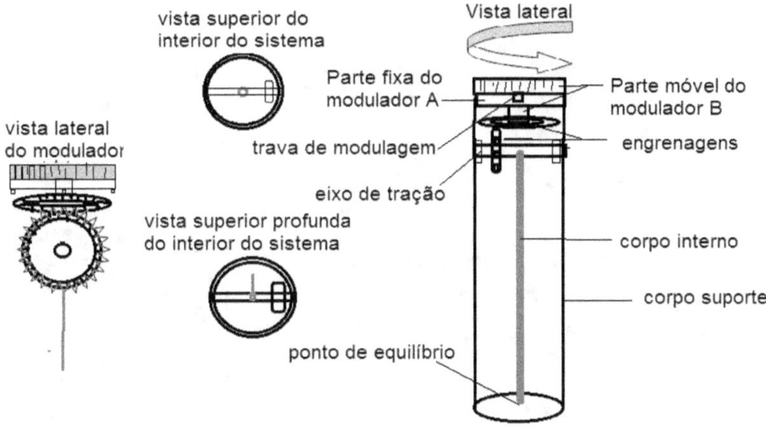

137

CORPO DIALÉTICO PUJANTE C

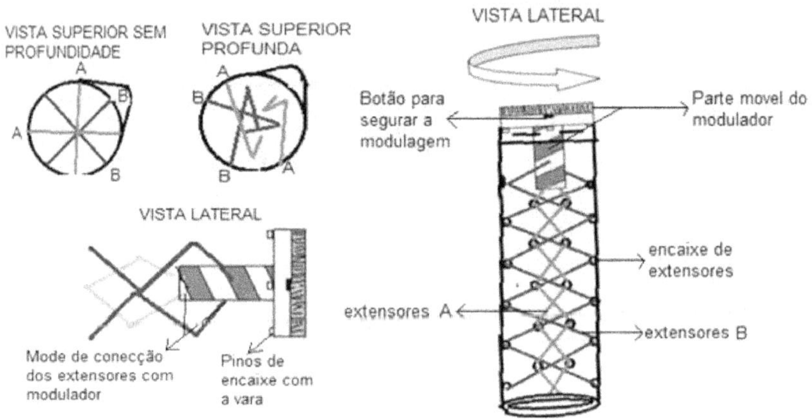

CORPO DIALÉTICO PUJANTE D

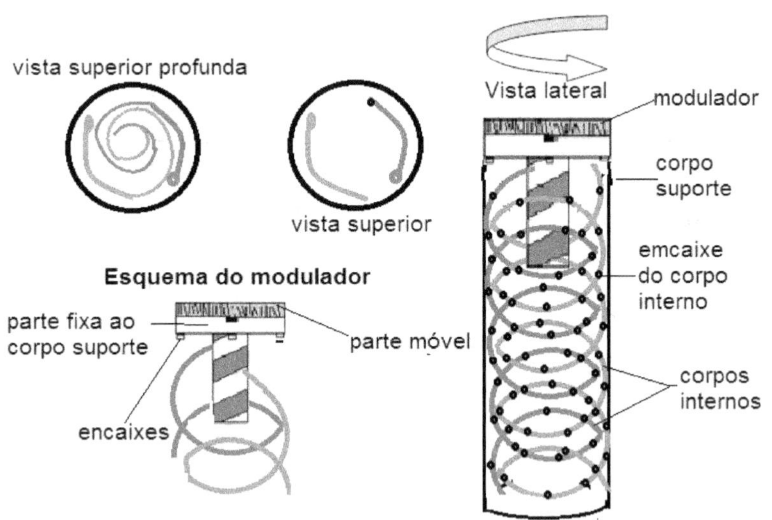

Source: author himself

138

FUTURE RESEARCH

The use of electromechanical modulation systems is advised for future research. As well as the detailed characterization of the first three types of systems.

For lines of technological research, it is advisable to use test vehicles for the springs of dialectical-powerful systems. Especially bicycle and motorcycles.

REFERENCES

MARTINS, W.C.; LAGO, W.W.S.; LOPES, J.S.; CUNHA, J.A.; SILVA, M.F.; The use of art and engineering to manufacture dialectical-thriving systems. Project and report, NPPGI, IFMA – Campus Zé Doca, partial report, Zé Doca, 2014.

BRITO, H.; Basic strength stroke of materials: pure tensile or compression. shear. Workbook of the Department of Engineering and Geotechnical Structures, University of São Paulo, 1st Issue, São Paulo, 2010.

139